MARS
UP CLOSE

NASA's Curiosity rover made a precision landing in Gale Crater during the summer of 2012, starting a breakthrough journey into once watery lowlands and then on to Mount Sharp. NASA's High Resolution Imaging Science Experiment (HiRISE) camera, orbiting Mars, captured the landing site (in blue) and surrounding scour marks created by retro-rockets fired for the final descent.

MARS
UP CLOSE

INSIDE THE CURIOSITY MISSION

MARC KAUFMAN

NATIONAL GEOGRAPHIC

WASHINGTON, DC

To my father, Irving Kaufman, who has always loved
books and taught me a deep appreciation of them.
And in memory of my mother, Mabel Kaufman, who
was an avid reader and learner as well.

CONTENTS

The Curiosity mission has discovered a new and different Mars—one where water was once abundant and the environment conducive to life. The breakthrough came at Yellowknife Bay, where Curiosity took this self-portrait.

Experience Mars
Up Close in 3D

Throughout the book, you will see icons like the one below and opposite. These are signals that, thanks to the digital engineers at NASA's Jet Propulsion Laboratory, allow you to have your own three-dimensional experience of Mars using your smartphone or tablet.

First you need to download NASA's free Spacecraft 3D app, available for Apple and Android at any app store. Once you have the app installed, point your smartphone or tablet camera at any image in the book accompanied by the icon. A 3-D model of one of the key Mars spacecraft will emerge from the image. Move your device around, move farther or closer, and (with the rover) operate the arm and pivot to create your own augmented reality experience of the spacecraft as they explore the planet Mars.

NASA

Jet Propulsion Laboratory | California Institute of Technology

SPACECRAFT 3D

AUGMENTED REALITY
View this image through NASA's Spacecraft 3D app for an up close experience of a Mars spacecraft.

CURIOSITY Watch from the JPL control room as Curiosity lands on Mars.

TRACK THE ROVER Follow along with this full traverse map, frequently updated by the Curiosity team.

The Red Planet isn't so red after all. Much of the Yellowknife Bay area contains flat, "platy" bedrock, as seen in this sandstone outcrop. The mixed light and dark layers reveal the actual color of the rock, a gray-black. The dark layers have been cleaned by the wind, while the lighter ones are covered in thick reddish dust.

FOREWORD by ELON MUSK

I n the next few decades, I plan to travel to Mars and make it my home. And if you're interested in having an experience like no other, I plan to someday make it possible for many of you to come as well.

This may sound like science fiction, but it's not. I've already had experience with setting goals in space that seemed unreachable, and now the company I founded to achieve those goals is carrying cargo to the International Space Station, and we're competing for the right to transport crew up there in the years ahead. Not long ago, sending a vehicle like Curiosity to Mars was science fiction, too. The pictures and discoveries coming back from the rover are our new reality.

Our efforts in space began a little over a decade ago. I had sold a profitable company and a friend asked what I was going to do next. I told him I'd always been interested in space exploration and travel, but it was an interest from a distance. There weren't clear ways that an individual could take on that challenge.

We talked a little further and agreed that the next big step for humans in space was Mars. People should be traveling to Mars, and doing it in our lifetime. I started exploring the Internet to learn what was already in the works. Other than NASA's long-term plans, there was nothing. I was shocked. It seemed crazy: an enormous opportunity largely ignored.

One thing led to another. It became clear to me that the real reason we weren't going to Mars was because our rocket technology cost too much. A 1989 estimate put the cost of a manned Mars mission at $500 billion, and that stopped the dreamers. So I set myself to the task of learning rocket science, meeting the people working in the field all around the world, and rethinking the way that rockets and capsules could be made.

Getting to Mars is too big an
accomplishment for us to feel
proud of just swinging by.
We are a nation of enterprise.

That's how SpaceX was born.

By bringing the cost of rocketry down—and inspiring trust in the private sector's role in space exploration along the way—we and others like us are bringing Mars closer. NASA remains the indispensable player, but it is time for new thinking on many different fronts. One way is designing reusable rockets, such as the ones we are developing in our line of Dragon spacecraft. Others ways are by reducing the weight of rockets without sacrificing strength and reformulating rocket fuels for greater bang for the buck.

Only by breaking through to new paradigms of space travel will more than a handful of us ever get to Mars and make it a potentially livable place. The United States landed men on the moon six times but then ended further exploration of it. I am talking about people settling on Mars and making life multiplanetary. It would be the start of a migration that would change our view of everything, parallel to Columbus or the Jamestown settlement and as momentous as the Wright brothers' first flight or Apollo 11's success with Neil Armstrong and Buzz Aldrin walking on the moon.

Getting to Mars is too big an accomplishment for us to feel proud of just swinging by. We are a nation of enterprise as well as exploration, and we're not about to go there without making something of it. When I think of going to Mars, I think of building greenhouses packed with rehydratable nutrients. I think of an iron-ore refinery. I think of a pizza joint.

This book fits right in with my sense of a mission to Mars. In all ways possible, it brings us closer to our neighbor planet. The pictures, whether taken by orbiting satellites or

No spaceship has ever landed on the Martian
moon Phobos, but some have tried. Futurists
describe it as a possible way station to Mars.
This artist's rendering imagines such a scenario.

SpaceX is in early development of private operations to travel to Mars and to establish colonies there—a long-term goal of company founder Elon Musk. Here an artist's rendering envisions the company's Dragon spacecraft on the planet.

<< SpaceX founder and CEO Elon Musk takes questions at the company's rocket development facility in McGregor, Texas. Behind him is the SpaceX Dragon capsule.

on the surface by the Curiosity rover, help us feel more at home with this landscape. The text brings us into the control rooms with those who have been masterminding this Mars mission for years. I have known the author, Marc Kaufman, as a probing science journalist who can see the big picture, and he has worked diligently to understand all phases of the Curiosity mission and the planet it explores.

You will visit Mars as never before in these pages. Hopefully the stories and pictures will help you imagine a day when your children, or their friends or children, will travel there. And maybe live there, too.

Sending large numbers of people to explore and settle Mars in the decades ahead isn't inevitable, but it is entirely possible. The biggest challenge isn't the engineering and spacecraft, however difficult they may be. Instead it's making sure that a sustained Mars campaign proceeds as a national priority, and that will happen only if the American people are behind it. We have the opportunity now to make this happen. We might not be so fortunate in the future.

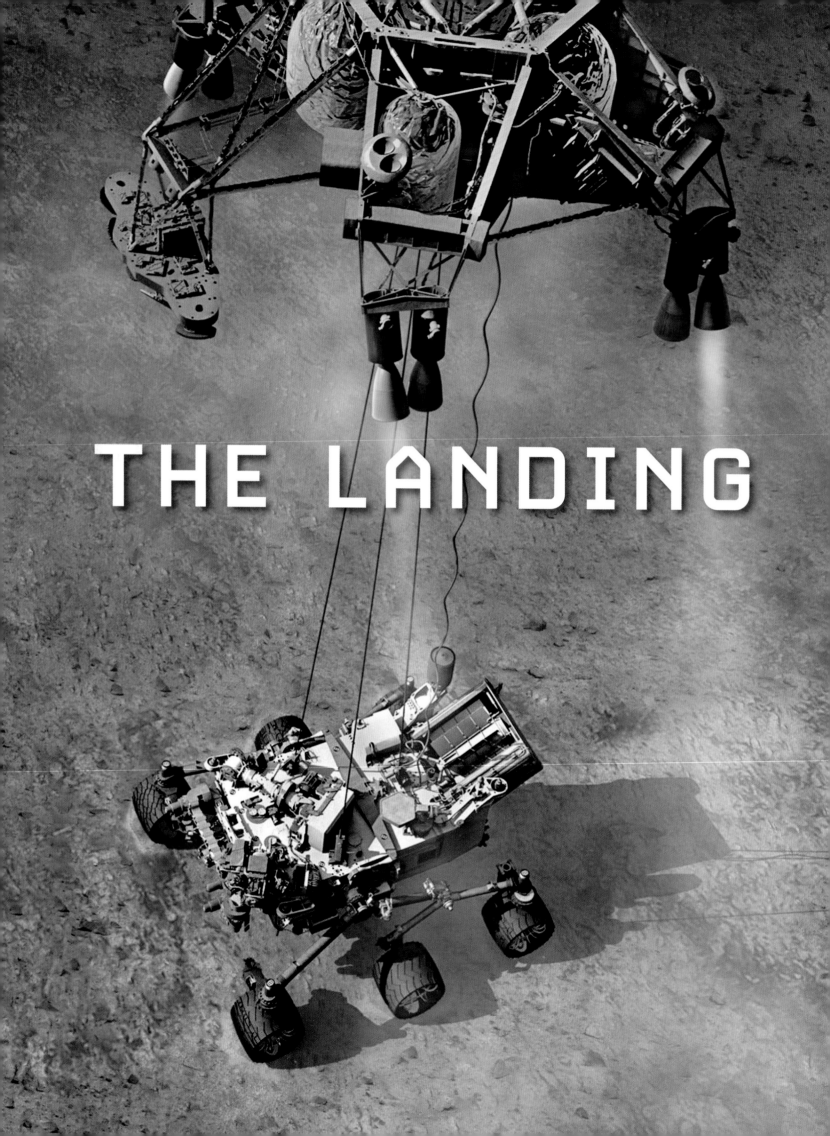

THE LANDING

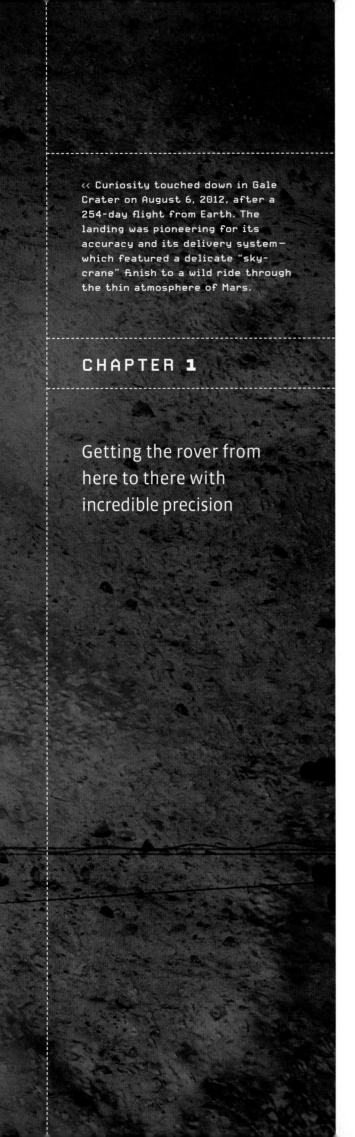

<< Curiosity touched down in Gale Crater on August 6, 2012, after a 254-day flight from Earth. The landing was pioneering for its accuracy and its delivery system—which featured a delicate "sky-crane" finish to a wild ride through the thin atmosphere of Mars.

CHAPTER 1

Getting the rover from here to there with incredible precision

IF YOU COME UP with a big new idea in our world and everyone says, "Hey, that's great, definitely go ahead with that," then you know it's not a big new idea at all. Anything really new brings out all the reasons why it can't possibly work, and why it's crazy even to think about it. That's certainly what we heard.

So explained Adam Steltzner, a former rock 'n' roll guitarist with the ear studs and Elvis pompadour to prove it. But in his professional guise, he was one of the NASA mission masterminds behind the most ambitious landing on another planet ever designed—the Curiosity touchdown on Mars.

He and his team at the agency's Jet Propulsion Laboratory, or JPL, had been tasked with the job of landing a one-ton, SUV-size rover on Mars. Their solution was to use a revolutionary, untried "sky crane." The capsule would hurtle down from space at 13,200 miles an hour to near zero in less than seven minutes. What remained of the entry spacecraft would use rockets to maintain a steady, 1.7-mile-an-hour descent.

Beginning some 65 feet (20 meters) above the floor of Gale Crater on Mars, the rover, dubbed Curiosity in a national naming competition, would then be gently lowered to the surface via three tethers and a central communications and power "umbilical." A risky venture for sure.

Following NASA tradition, Steltzner and his sky-crane colleagues decided to name their sky-crane landing as well.

Failure, as was famously said of the crippled
Apollo 13 spacecraft, was not an option.

They considered their plan to be audacious and relished the fact that they were willingly taking on the challenges of Mars in a new way. The name wasn't hard to agree on: The Entry, Descent, and Landing (EDL) phase would be called Audacity.

As the August 5, 2012, landing day approached, the indefatigable Steltzner had become the articulate and emotionally accessible, very worried, but so clearly talented face of the touchdown drama—the upcoming "Seven Minutes of Terror" that had captured the imagination of millions of people around the world.

This rendezvous with Mars was one of numerous NASA missions to the planet, but the future of NASA's entire Mars program nonetheless was very much in the balance on landing night—an additional weight that Steltzner and his colleagues always carried.

After two missions to the planet failed in the late 1990s—during a time when NASA's operating philosophy was "faster, better, cheaper"—a new team was brought in to design a ten-year plan. They engineered a series of successful Mars missions with landers, rovers, and orbiting satellites, but future planning had stalled. The Mars Science Laboratory mission (or, less formally, the Curiosity mission) had been approved before the money began to dry up, and at $2.6 billion it was among the most expensive Mars ventures ever. Failure, as was famously said of the crippled Apollo 13 spacecraft, was not an option.

But success would not come easy. Curiosity was much larger and heavier than past rovers, carrying ten instruments rather than five, including two miniaturized chemistry labs. A heavier rover is a rover that's inherently more difficult to land safely. What's more, NASA wanted a much greater level of precision in terms of where the rover would touch down—hitting within an ellipse 15.5 miles (25 kilometers) long rather than the 50-mile (80-kilometer) range given the previous rovers, Spirit and Opportunity.

And then there's that peculiar, huge mound at its center, with enticing layers of rock that appeared to record the history of Mars. Getting to Mount Sharp was a key goal of the mission. Landing anywhere but inside the crater would make that goal unachievable, since the rover couldn't make it down the steep crater walls.

It was this challenge—landing a large and heavy vehicle on Mars with a greater accuracy than ever before—that Steltzner's team had tackled. The team leader knew they had solved all the thousands of engineering puzzles thrown at them; they had tested and retested their solutions many times over. In fact, in JPL speak, they had crushed the forces that would seek to make the landing fail.

Curiosity touched down within its planned landing ellipse—an intensively studied, flat target area much smaller than those for any previous Mars spacecraft. While the Curiosity landing broke all records for accuracy, it remains far from the level of touchdown precision needed to land astronauts on the planet.

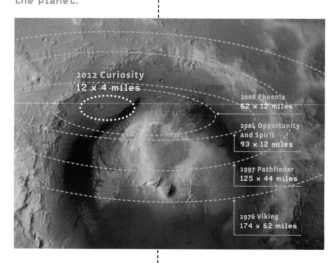

2012 Curiosity
12 x 4 miles

2008 Phoenix
62 x 12 miles

2004 Opportunity
and Spirit
93 x 12 miles

1997 Pathfinder
125 x 44 miles

1976 Viking
174 x 62 miles

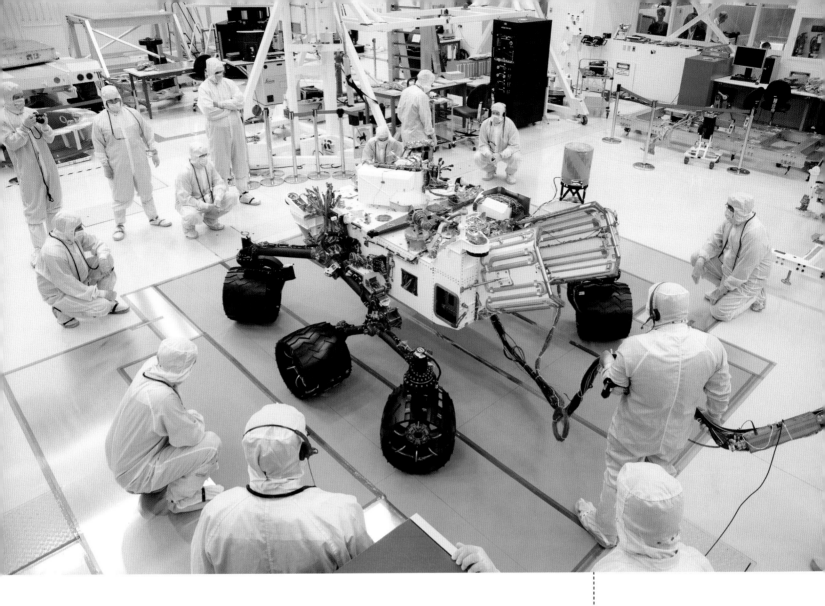

So Steltzner's intellect told him there was every reason to be confident. Yet beneath that conscious, fact-based optimism and outward calm was something quite different. He worried about the parachute, an enormous supersonic brake that came with known risks and a definable chance of failure. He worried about the "guided entry" flight path, controlled by an autonomous system inside the capsule that would produce a series of essential S turns. Guided entry had been used by the Apollo astronauts on their moon and Earth landings, but hadn't been employed successfully since.

And he worried about the sky crane that he and a handful of others had designed and championed. Eight retro-rockets would have to begin firing just after the stage separated from the parachute, slowing what remained of the spacecraft (the sky crane and a rover) to a near hover above the ground. Slowly, slowly, the precious cargo would be lowered to the planet's surface. At touchdown, the tethers would snap and the sky crane would fly off to crash some distance away—unless, of course, the jets didn't get the fly-away message and came crashing down onto Curiosity.

He worried about these big-picture risks and more, but what truly terrified him was something different. It was the "unknown unknowns" out there, the hidden problems that nobody knew to even worry about. That's what gnawed at him constantly. That's why Steltzner would later say that he arrived for landing day feeling "emotional terror" and that he was in a "cauldron of despair."

Definitely not the NASA speak of missions past.

Engineers and technicians in clean-room garb monitored the rover's initial drive test in 2010. A first-of-its-kind rover, new technologies and systems were needed throughout its design, development, and construction. Curiosity was assembled at NASA's Jet Propulsion Laboratory in Pasadena, which is managed by the California Institute of Technology.

MISSION MAKERS >> ADAM STELTZNER

"Exploring is fundamentally human; we've done it for thousands of years. It's an expression of something that's the best in us."

Born an heir to the fast-dwindling Schilling spice fortune, Steltzner almost flunked out of high school and did flunk high school math (for skipping so many classes). But at 21, after finishing a nighttime gig with a band called Stick Figures, he was puzzled by the apparent celestial movements of the constellation Orion. That newborn interest in astronomy led to a physics class at a local community college, a stellar undergrad career at the University of California, Davis, a master's at Caltech, and a doctorate from the University of Wisconsin. He came to JPL in 1999 and helped design the inflatable balls that delivered the rovers Spirit and Opportunity to the surface of Mars.

Jet Propulsion Laboratory, NASA

The lead engineer for Curiosity's Entry, Descent, and Landing, Steltzner advocated the pioneering "sky-crane" landing, including a parachute designed to withstand supersonic speeds, tested (below) at NASA's Ames Research Center.

Just as Steltzner brought a brash, emotional, and intriguing new face to NASA, so too did Curiosity change the look and reach of planetary exploration. Virtually everything about the mission pushed the envelope: its ambitions, its designs, its destination, its cost.

Its scheduled 23-month expedition (expected to actually last much longer) was formally and officially a "mission of discovery," free to follow the scientific leads that materialized. But Curiosity was also tasked to test and expand on a revolutionary new understanding of Mars that has emerged in recent years. Satellite photography and orbital readings of minerals and compounds on the Martian surface have convinced many scientists that Mars was wet and warmer in its distant past; that flowing rivers, streams, and lakes were common; and that minerals were deposited there that, on Earth, are associated with life. From above, the planet has been looking increasingly like a once habitable place.

But detecting all this from orbit is one thing; actually conducting the rigorous surface experiments needed to confirm, modify, or discard this evolving view is quite another. Mars scientists call the process "ground-truthing," and Curiosity's chemistry labs, its laser blaster, its first ever drill, its ultra-high-resolution cameras, and more are the kinds of instruments that can provide some answers.

Hundreds of scientists and engineers studied and culled a long list of candidate sites before selecting Gale Crater as the most promising. Most enticing was the presence of Mount Sharp, with its deep canyons, mineral deposits, and riverbeds. Three miles (five kilometers) high, it promised an unprecedented geological reading of the planet's history as laid down in exposed layers of rock. The rover has a nuclear-powered battery that can and will power its explorations for up to 14 years if other systems remain healthy.

The expedition's ambitious and potentially very long-term plan, then, was to explore Mars with the primary goals of finding habitats where living organisms could have once survived and understanding the global conditions those organisms would have faced. Proving that early Mars had been habitable would be one of the signal accomplishments of the space age. But it would also throw open the door for future missions focused more specifically, and more knowledgeably, on whether life actually ever did arise on Mars.

So much scientific adventure and excitement. But first Curiosity had to reach Mars and land safely inside one of its craters.

LIFTOFF

Blastoff from Cape Canaveral lit the early-morning sky on November 26, 2011. Lifting a one-ton vehicle (and three-ton spacecraft)

A central goal of Curiosity is to explore sedimentary Mars, a part of the planet where layers of broken-up lava and other material formed into rock. McLaughlin Crater is one of many sedimentary sites.

CLAY-CARBONATE LAYER SURVEY SITES

Getting the launch right didn't ensure

a successful landing.

into space is no small feat, and it had to be done so the spacecraft was almost immediately flying in the right direction, with the right trajectory, and at the right speed.

Getting the launch right didn't ensure a successful landing, but getting it even a bit wrong would reduce the odds of success dramatically. So the launch team collaborated broadly with the flight and landing teams. More than eight years in development (including an initially controversial and costly two-year delay to complete some unfinished work), the mission would spend $2.6 billion to land America's seventh vehicle on Mars. Other nations have tried Mars landings, but only NASA's have operated successfully on the planet.

The launch crowd included many from the JPL flight, landing, and science teams, and that meant Steltzner and his wife, Trisha. As fate would have it, the two were about to start another journey of their own, and it was on a nearly identical timetable. It wasn't planned per se, but Trisha was about to become pregnant, with a due date right as Curiosity would be approaching Mars nine months later. Dramas within dramas.

Many of the other Curiosity engineers in the crowd had, like Steltzner, worked on the Spirit and Opportunity rovers. In 2004, the two were dropped onto the surface of Mars inside two large balls. When they landed, the "little rovers that could" had a life expectancy of 90 Martian days, or sols. That's a little more than three months. But Spirit lasted until 2009 and Opportunity, as of late 2013, was still exploring and making important discoveries. Much smaller and less sophisticated than Curiosity, they had nonetheless paved the way for their rover sister-to-be.

Others had worked on the Phoenix lander, which in 2008 had touched down in the northern polar region of Mars and had dug down a short distance and found water ice. A twin of one of the earlier spacecraft that failed, its successful mission substantially expanded the playing field in terms of possible landing sites, and made scientific findings that would have great significance for Curiosity.

And some in the crowd were responsible for two highly successful satellite monitoring stations in orbit around the planet. The Mars Odyssey arrived in late 2001 and carries a unique instrument for detecting minerals on Mars, particularly those that can only form in the presence of water. The Mars Reconnaissance Orbiter was launched in 2005, and its High Resolution Imaging Science Experiment (HiRISE) camera has done more to transform our understanding of Mars than perhaps any other instrument sent to the planet.

So there was a lot of experience, a lot of confidence, a lot of reason for confidence. All the myriad parts of the Curiosity mission were tested and retested, modified and thrown out. But there was one test that could not be conducted, and it was a big one.

Steltzner and the rest of the team would be landing with the painful awareness of this sobering truth: They would be conducting their first actual "trial run" of the full and enormously complex system in the atmosphere and on the surface of Mars.

BLASTING OFF ›› NASA's Mars Science Laboratory, which carried the Curiosity rover inside, sits atop an almost 200-foot-high Atlas V rocket (below) in the days before takeoff from the Cape Canaveral Air Force Station. After launch on November 26, 2011 (above, right), the rocket released the spacecraft into the upper atmosphere (artist's rendering, above, center). Technicians worked on the program—which was more than nine years in development, including a two-and-a half-year delay—at NASA's Payload Hazardous Servicing Facility at the Kennedy Space Center (above, left).

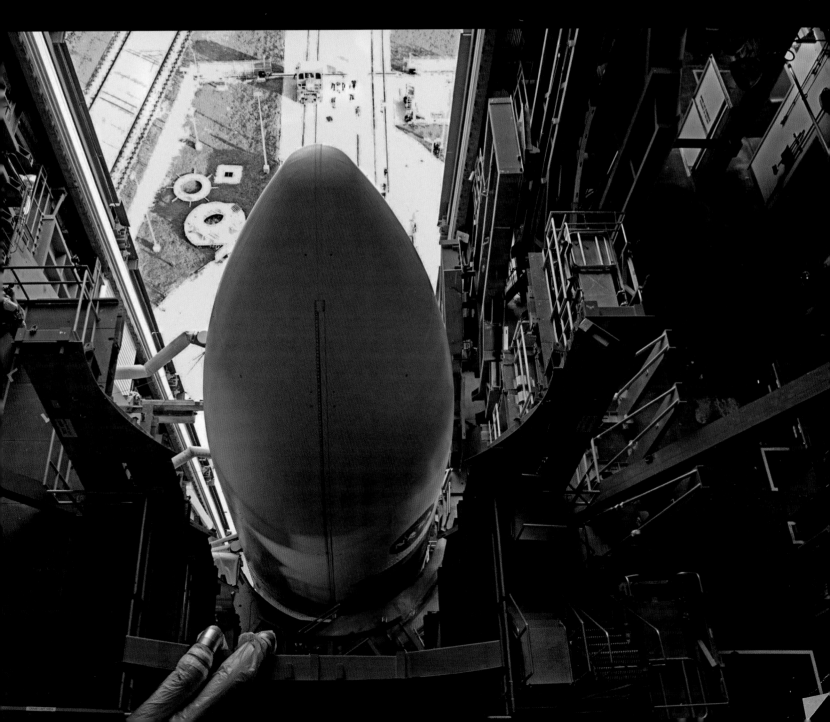

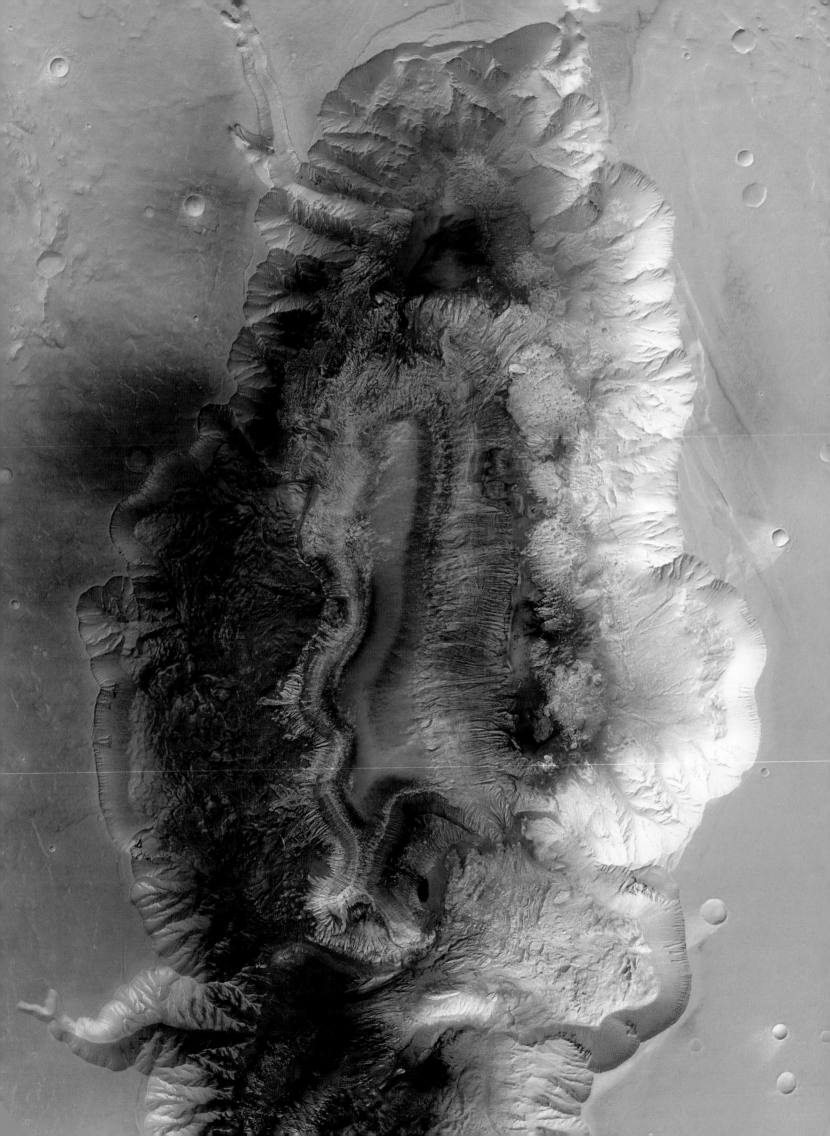

NO MORE WAITING

Landing day at the Jet Propulsion Lab campus on the outskirts of Pasadena was both excited and tense in the extreme. It was, ironically, also a time of some sadness for the Entry, Descent, and Landing team, which had worked so long and hard together and now was about to break apart—whatever happened. As a final pre-event hurrah, Steltzner took the 50-member EDL team out for a lunch at the Burger Continental restaurant in Pasadena, a Greek restaurant complete with belly dancers. The location was deemed to hold good luck, since Steltzner had taken an earlier rover landing team there, too.

That evening, with the landing only a few hours away, Steltzner brought his crew to the central mall of the JPL campus and presented them to the growing crowd, which cheered wildly as if the hometown football team had just arrived. All in matching light-blue polo shirts, the EDLers assembled before a model of Curiosity for a grand team photo. When the flashes stopped, they marched to the stations at Mission Control and the nearby "war room," where they would monitor—and, to a limited extent, direct—the landing.

At that point, Curiosity and its cruise vehicle had traveled some 352 million miles (566 million kilometers) to reach and synch with its orbiting target. The flight path established soon after launch was so good that the JPL team needed to make only small tweaks over the ensuing 254 days in space. The three big Earth-based dishes of the Deep Space Network were sending back position and speed, and the spacecraft was making its final approach right on target.

At 10:17 p.m. Pacific time on August 5, 2012, the landing craft began its separation from the cruise spaceship. Controlled pyrotechnic explosions announced the onset of the detachment as the systems of the two vehicles were precisely but forcefully blown apart.

Among the small explosions were those that set three miniature guillotines into action, slitting the cable and wiring connections. Curiosity was still ten minutes from actually hitting the top of the Martian atmosphere, and it was newly on its own.

Those feared *unknown unknowns* no longer mattered. The spacecraft would land safely or crash in a ball of fire on its own.

AN INSIDER'S VIEW

Hundreds of men and women worked for years on the Curiosity landing, and so Steltzner had resisted being singled out as the voice and face of what was about to happen. Nonetheless, he was the lead of the EDL team and at the center of the night's drama.

Only six people from the EDL team were actually in Mission Control for the landing; the room instead was packed with an operations team of engineers and project managers, as well as some top NASA and JPL officials.

On one side of Steltzner was Miguel San Martin, lead for Guidance, Navigation, and Control, and by all accounts another mastermind of the landing. On the other side was Al Chen, who did the play-by-play for NASA TV.

That version was seen and heard by millions at "landing parties" around the world and around the clock—from Times Square to St. Petersburg, from Rome to Sydney and Singapore.

It was thrilling, the
end of eight years of
working and waiting.

NASA Administrator Charles Bolden (above)
waits in prayerful silence on landing night.
Agency Science Mission Director John
Grunsfeld (right) was at JPL Mission Control
as well. Both Bolden and Grunsfeld are former
astronauts but had much riding on this most
ambitious robotic mission to Mars.

But the play-by-play inside the heart and mind of Adam Steltzner was equally compelling
and is known to far fewer. It went something like this:

Click . . . click . . . click.

*Steltzner was suddenly a boy again and on a roller coaster inching up the incline to the
top of the ride.*

Click . . . click.

*The ten minutes between separation and Curiosity's first contact with the Martian atmo-
sphere was endless, as if time froze. Steltzner was feeling it: There's nothing I can do to stop the
climb and the wild ride to come. Am I ready for this?*

*The top of the climb, of the ride, came into sight, into the mind's eye. It was thrilling, the
end of eight years of working and waiting.*

*But it was overwhelming, too, an emotional jumble. What if the roller coaster flew off the
tracks? What if the spacecraft crashed and burned?*

*To make everything a touch more surreal and unnerving, if possible, Steltzner knew that
by the time word came that Curiosity had entered the Martian atmosphere, it actually would*

have already safely landed or crashed. That's because of the almost 14-minute time delay in data streaming from Mars—the time it takes for signals to travel the millions of miles from the planet to Earth.

The suspense was no longer what would happen, but what *had* happened. How weird was that?

The clues would come first from MEDLI, the Mars Science Laboratory Entry Descent and Landing Instrument. Being used for the first time on a landing mission, the sensing instrument contained an array of pressure tubes and thermal couplings that would tell NASA how the heat shield was performing. Since the shield would have to withstand temperatures up to 3800°F (2100°C), NASA wanted to know all it could about its strengths and weaknesses for the descent in progress and those that will follow. But for MEDLI to work right, it had to remain at a steady -20°F (-29°C); no small feat.

The ten minutes passed and the roller coaster, they had to assume, began its plunge down. Standing between Chen and San Martin, Steltzner glanced over to Chen's monitor, which would be getting the first readout from MEDLI via the Deep Space Network.

PREPARE FOR LANDING >> In this artist's rendering of the Mars Space Laboratory (MSL) approaching Mars, the spacecraft's cruise phase has been jettisoned, a step that occurred about ten minutes before entering the atmosphere. At that point, the spacecraft was 2,188 miles (3,522 kilometers) from the center of the planet.

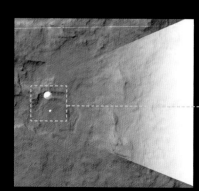

PARACHUTE OPENS The high-resolution camera on the Mars Reconnaissance Orbiter captured the brief moment when Curiosity's deployed parachute was slowing the spacecraft down for landing.

What he saw was this message: "Beta out of balance. Catastrophic."

Whoa.

What that signal meant, if it wasn't a calibration or transmission error, was that the descent capsule was slipping sideways—yawing—much more than called for, and there would be major trouble ahead.

Chen turned to Steltzner, tapped him urgently, and pointed to the message Steltzner was already reading. Their eyes met.

What the hell?

Let's hold this together, keep it easy. Let's see if this is real . . .

Those unknown unknowns. The seconds ticked by. Nobody has to know this . . . yet.

Then the message changed as quickly as it had appeared. It had not been real. A steady heartbeat pulse from MEDLI appeared, and suddenly the approaching disaster vanished as quickly as it had come into view.

The message turned out to be a calibration error made during testing on Earth, and no harm was done. Except, of course, to the already seriously frayed nerves of the two people whose introduction to the actual entry and landing contained the word *catastrophic*.

SAVORING "NORMAL"

But soon Steltzner was receiving a steady stream of good news in his ear. If reports come in that some maneuver, some firing, some important event happens as planned, the accomplishment is passed on as having been "nominal." No problem.

The thrusters that stop the spacecraft from spinning at two rotations per minute fired off as planned. Nominal.

More thrusters fired on schedule to push forward the capsule's heat shield, a key moment called "turn to entry." Nominal.

Then two tungsten weights (each 165 pounds, or 75 kilograms), used for balance while flying to Mars, were jettisoned, tipping the spacecraft and its heat shield up a bit so they would have more lift. Nominal.

The self-adjusting guided-entry system needs that lift so it can surf variations in the Martian atmosphere and steer a more precise course, using a series of S curves in the upper atmosphere. A very big deal.

Steltzner savored the word. *Nominal, nominal.* He finally was feeling a little bit good. On edge, but good. And the higher-end com was starting to flow down, giving a strong and clear signal.

"Com" is communication, and it was no longer coming through the low-bandwidth transmission from MSL directly to the Deep Space Network, but rather via a relay from NASA's Mars Odyssey satellite. Odyssey was one of three orbiting the planet and positioned and primed to monitor—and even photograph—the Curiosity descent. Each important

HEAT SHIELD The MSL shield, the largest ever built for a planetary mission, protected the spacecraft as it passed through the Martian atmosphere (artist's conception, right).

> The possibility of success entered
> Steltzner's mind. Yes, yes, yes, we're
> going to do it. Then he was yanked back.

activation on the way down would be announced by a series of specific, very basic radio tones. One sound would say the parachute deployed, another would signal that the airbags had inflated: a succession of 128 individual beeps or tones to announce the downward progress.

After 80 seconds in the Martian atmosphere, friction on the heat shield had pushed the temperature up to 3800°F (2100°C), slowing the capsule to 10 percent of its original speed. But since it hit the atmosphere at 13,200 miles an hour, it still needed substantial braking.

Now it was time for the parachute to open—one of the more unpredictable stages, where some risk was inevitable. Steltzner wasn't focused so much on that opening, but rather on the eight rockets that would be needed for the final sky-crane landing.

The rockets had to be opened to full throttle as the parachute deployed to keep them from rattling and interfering with the aerodynamics of the maneuver. Then they had to return to their 2 percent open position for the next stage. And now they were back at 2 percent, and a dangerous hurdle had been cleared.

Precision work, real fine precision work. The retros were nominal once more, a huge relief.

The capsule was speeding down, really ripping, and everything looked very good.

Fifteen voices were in Steltzner's ear, reporting in. But people were stepping over each other, and he was calling out for repeats.

The possibility of success entered Steltzner's mind. *Yes, yes, yes, we're going to do it.*

Then he was yanked back.

Getting totally gleeful led to getting totally anxious. It was like putting together a house of cards: getting higher and higher and reaching the point where you'll either make it happen or it will all come crashing down.

Adam Steltzner, the near embodiment of cool, was about ready to lose it.

But the parachute did its job, the descent stage pulled away, and the sky crane with its precious cargo was soon some 65 feet (20 meters) above Mars—ready to initiate the final landing maneuver. For Steltzner in particular, this was the climatic moment—not only because it was a high-risk and untried landing, but also because he was so completely identified with it. A dozen others made key contributions, and San Martin was the one who figured out how to implement it all. But Steltzner had also made key technical contributions, and, more important, he had sold it to the powers that be.

So while there were more than a few coparents holding their breath as their proudest creation went into action, there was really only one leading midwife—Steltzner. If something went bad at this last stage, he knew, he would get no sympathy at all. He and his fellow EDLers had pushed too hard for that; it was their baby.

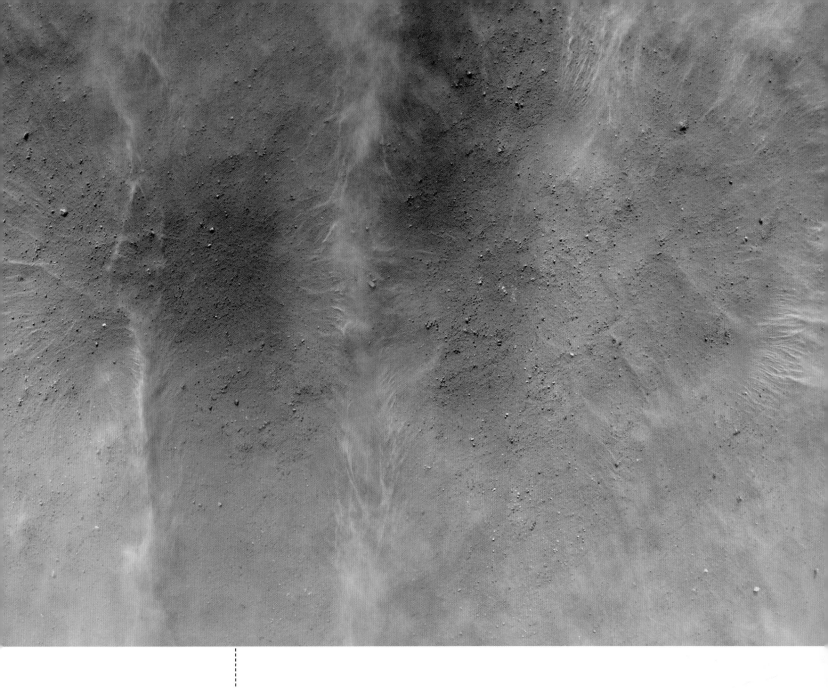

Touchdown kicked up a swirling cloud of dust, captured by the Mars Descent Imager (MARDI). The camera was placed on the underside of the rover, and it took three minutes of the first ever images of a Mars descent.

HAS CURIOSITY LANDED?

By the time Al Chen was reporting that the sky crane was in action, Steltzner had already known it for a few moments. In fact, he had been getting a more technical stream of information all during the descent.

That's because the team had decided Chen could announce a maneuver as having been accomplished—the heat shield detaching, the ballast being dropped—when he got word that the capsule had sent out the commands to take that action.

But confirmation of the actual fulfillment of that order came in a different stream of data that was far too complex for the play-by-play. If the command announced by Chen somehow had not been followed, Mission Control (and the world) would know soon enough: The progression of commands from the capsule's computer would simply stop.

There was one exception to this rule: a successful touchdown. That had to be confirmed by data, not just a command from the capsule computer, and it had to be confirmed in three different ways. No way was Steltzner going to allow that news before it was absolutely, completely confirmed. No way was he going to risk having to walk that one back.

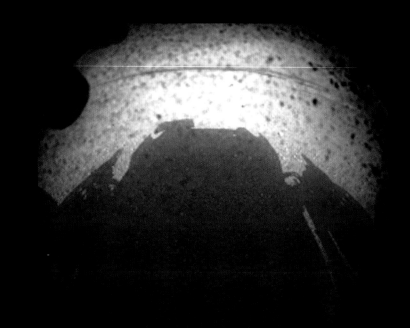

VERY FIRST PICTURES >> The successful touchdown on the Mars morning of August 5 or August 6, 2012, on Earth (depending on the time zone) set off jubilation at JPL and also at landing parties from Times Square to Seattle and from Adelaide, Australia, to Tbilisi, Georgia. In Mission Control, the joy was reprised (above) when the first image came back from Curiosity. The picture on the Mission Control screen was taken soon after landing by one of the front hazard-avoidance cameras (Hazcams) attached low on the rover chassis.

Both the front- and rear-left Hazcams sent their fish-eye images almost immediately. In this early front Hazcam photo (left), the silhouette of the rover makes up the bottom of the image and the Martian surface can be seen on top. The camera's dust cover remains in place, and Mars dust can be seen around its edge, along with three cover fasteners.

> The sky crane was happening, and all he could think
> of was the ways it could die. He was waiting, waiting.

So before Chen could announce success, this is what had to happen:

First they needed proof of the six big aluminum wheels on the ground. The rover produces an "event record" that would report contact (via an indirect measure from the descent stage) in real time via the Odyssey satellite—or real time plus the 13-minute, 48-second lag from Mars to Earth. It was the job of Jody Davis, of NASA's Langley Research Center, to make the call.

But Davis would not be shouting out, "Touchdown." Steltzner and others didn't want anyone in Mission Control or Televisionland to hear "touchdown" before the other two signals came through. So her message of a successful wheels-down would be "Tango Delta."

Next came a signal that would describe whether the rover was in motion on Mars. Landing on a slope, a dune, or even the crater wall could mean big trouble, and so the second call would be whether the Rover Inertial Measurement Unit—RIMU—was reporting motion (a bad thing), and whether Curiosity was generally oriented as expected. Dave Way, also of Langley, would be monitoring the RIMU in the war room.

Finally, the descent capsule shell, the one over Curiosity as the lines unspooled and the rover slowly dropped, had to immediately disconnect and fly away. If that didn't happen, it could well crash onto the rover below, with truly catastrophic results.

That cut-and-fly maneuver was programmed to occur within ten seconds of wheels-on-ground. So if Curiosity was still sending back UHF signals via the Deep Space Network after that time, then all was well. If the signal stopped—well, that was the end of it. Brian Schwartz, an EDL communications engineer, would be tracking the UHF signal.

Back at Mission Control, Steltzner was somewhere between elation and craziness.

The sky crane was happening, and all he could think of was the ways it could die. He was waiting, waiting, and then he heard from Jody Davis: "Tango Delta nominal."

That was good, very good. Next, do we have RIMU? Yes! Dave Way called out, "RIMU stable."

Ten more seconds. Steltzner paced and counted.

Four, three, two, one. And then he looked over at Brian Schwartz.

UHF strong!

The descent shell was speeding off and wouldn't crash onto the rover.

Steltzner tapped Chen, who then delivered the good news. "Touchdown confirmed."

What followed was bedlam—among the most raucous, exuberant, we-are-the-champions moments in JPL history. Their job done, the entire EDL team piled out of the war room and Mission Control to, in effect, crash the post-landing press conference, everyone in perpetual high-five mode. It was past 1 a.m. in New York when the rover landed, but a crowded Times Square erupted with whoops, chants, and laughter. It was the same from Adelaide, Australia, to Tbilisi, Georgia, from Rome to Toronto, and from Austin to Tucson to Seattle to Cleveland to Yellowstone National Park.

Curiosity was safely on Mars.

CURIOSITY GOES TO MARS

BLASTOFF An Atlas V rocket sends the spacecraft carrying Curiosity on its 352-million-mile journey to Mars. The voyage takes nearly nine months.

TARGET IN SITE Spacecraft containing rover as it nears its destination

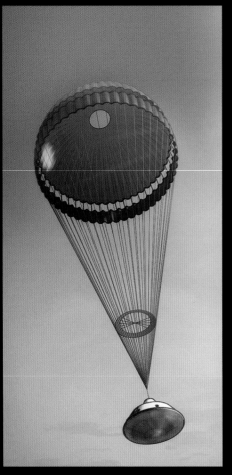

THE BRAKES The parachute deploys, slowing the descent.

SLOWDOWN The heat shield pops off, and the sky crane is released.

TOUCHDOWN Sixty feet above the surface, the sky crane lowers the rover

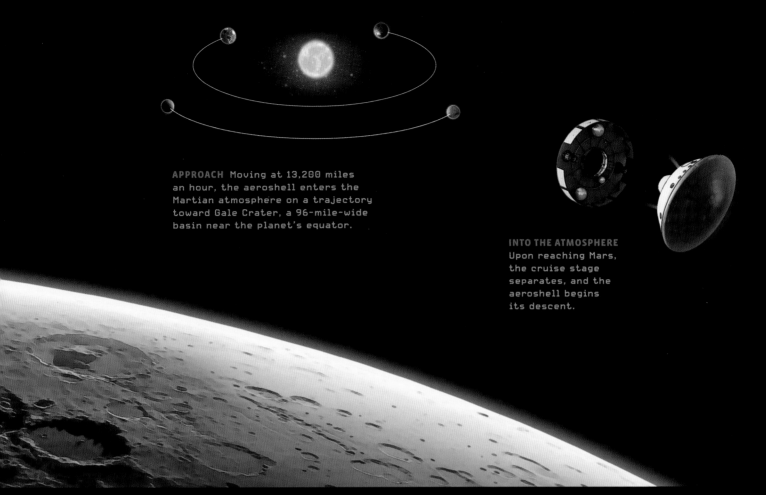

APPROACH Moving at 13,200 miles an hour, the aeroshell enters the Martian atmosphere on a trajectory toward Gale Crater, a 96-mile-wide basin near the planet's equator.

INTO THE ATMOSPHERE Upon reaching Mars, the cruise stage separates, and the aeroshell begins its descent.

LAUNCH November 25, 2011
LANDING August 6 (UT), 2012

AND FINALLY About the size of a small SUV, the rover Curiosity will search Gale Crater for past or present conditions favorable for sustaining life. Maximum speed: 470 feet an hour.

ON BRADBURY RISE

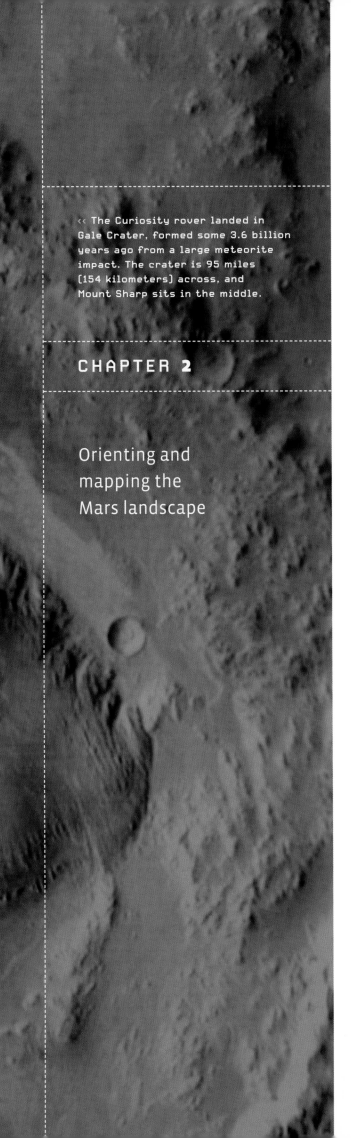

<< The Curiosity rover landed in Gale Crater, formed some 3.6 billion years ago from a large meteorite impact. The crater is 95 miles (154 kilometers) across, and Mount Sharp sits in the middle.

CHAPTER 2

Orienting and mapping the Mars landscape

THE FINAL DESCENT was gentle by Mars landing standards: no bouncing balls with rovers inside, no retro rockets at touchdown, no crashing and burning (the most common arrival on Mars). But still, the jets on the sky crane burned two scours in the ground and kicked up high plumes of the omnipresent reddish Martian dust, a caustic and constant reminder that life-limiting iron oxide compounds now cover the planet.

Unexpected golf ball-size rocks were also on the move—some at up to 170 meters a second (about 380 miles an hour) as touchdown approached. Curiosity was engineered to land with its equipment in a protective tuck, and only that design and some luck kept those speeding rocks and pebbles from doing real harm. Total damage was one broken wind sensor.

As the landing celebrations died down on Earth, the Curiosity team was already far along in answering the next pressing question: Where was the rover?

The answer was almost too good to be true, because it touched down inside the landing ellipse NASA aimed for, an area smaller than the size of Wichita Falls, Texas (4 by 12 miles). After 352 million miles of elliptical travel, Curiosity landed safely inside that smallest Martian landing zone ever—a journey likened in its precision to a baseball thrown from Atlanta and landing as planned in the glove of a fan in Dodger Stadium in Los Angeles.

BEFORE AND AFTER Early Hazcam photos show Mount Sharp in the distance, photographed with the camera's dust cover on (left) and off (right).

Chances are good that the rover will still be in the crater when the time of humans has ended on Earth.

The landing site was a rise several meters above the surrounding flatland of Gale Crater, Curiosity's new and now forever home. More than 95 miles (154 kilometers) in diameter and believed to be about 3.6 billion years old, it lies near what is called the Martian "dichotomy," the line near the equator that separates the endlessly flat northern plains from the southern highlands, river valleys, and volcanoes.

Chances are good that the rover will still be in the crater when the time of humans has ended on Earth. It certainly won't be in working order, but it will likely be there. Mars is a harsh place for people, but much kinder than Earth on inanimate objects: The only real threat to its physical integrity is a direct hit from a meteorite or comet that pulverizes it and everything around.

Curiosity is generally called a rover, but really it's a robot. In the days ahead, all its myriad commands would be tested and verified and, over the coming months, put to work. Included in those commands are the ones that tell the rover to shut itself off when something seems

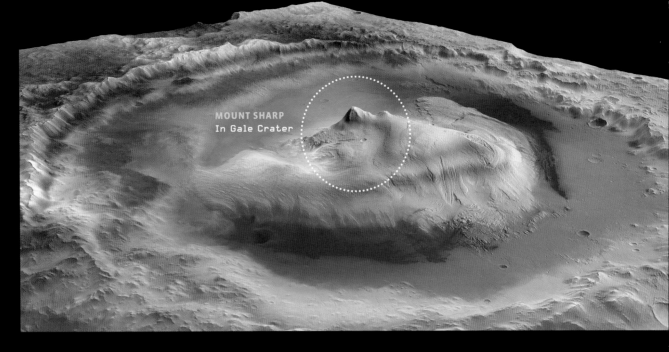

MOUNT SHARP
In Gale Crater

THE VIEW FROM CURIOSITY >> A 1.3-billion-pixel landscape view with Mount Sharp in the distance (opposite) shows one stretch of a 360-degree panoramic mosaic, made up of 900 images taken by the rover's camera and stitched together. A colored elevation map of Gale (above) shows the low area, shaded blue, in which the rover landed.

wrong, or to drive and make its own decisions about whether it's safe to move ahead. If not, the robot-rover is capable of making a new path to get where it's supposed to go.

Previous Mars rovers have taken on personalities of sorts, the inevitable anthropomorphizing that goes with that robot nature. As for Curiosity, it landed as a brainy, perky, complex, and almost female presence, a solitary and maybe a heroic stranger in a strange land. Soon it would awaken and show its mettle in a new and alien world.

Or was it so alien?

As the first images from the rover's black-and-white navigation cameras arrived at JPL, a common reaction from the scientists was that they were looking at a scene that looked surprisingly familiar. It was harsh and desolate for sure, but the crater featured cliffs in the distance, a changeable terrain that appeared to be made up of different types of rock layers and deposits, and that looming mountain was off to the side. What's more, the bedrock was covered in gravel and some larger rocks, not just sand and dust, and soon layered outcrops of rock came into view.

THE CRATER SITE

Although past landing sites were all surrounded by broad, flat plains, Gale Crater was, relatively speaking, a scene of geologic and landscape bounty. It looked rather like Death Valley, some thought. Or Chile's high Atacama Desert. Or maybe the McMurdo Dry Valleys of Antarctica. That was the choice of geologist Bernard Hallet, an analyst for the Curiosity camera team who specializes in cold and arid places: "Really, I looked at the scene and nothing looked unfamiliar from our work down there."

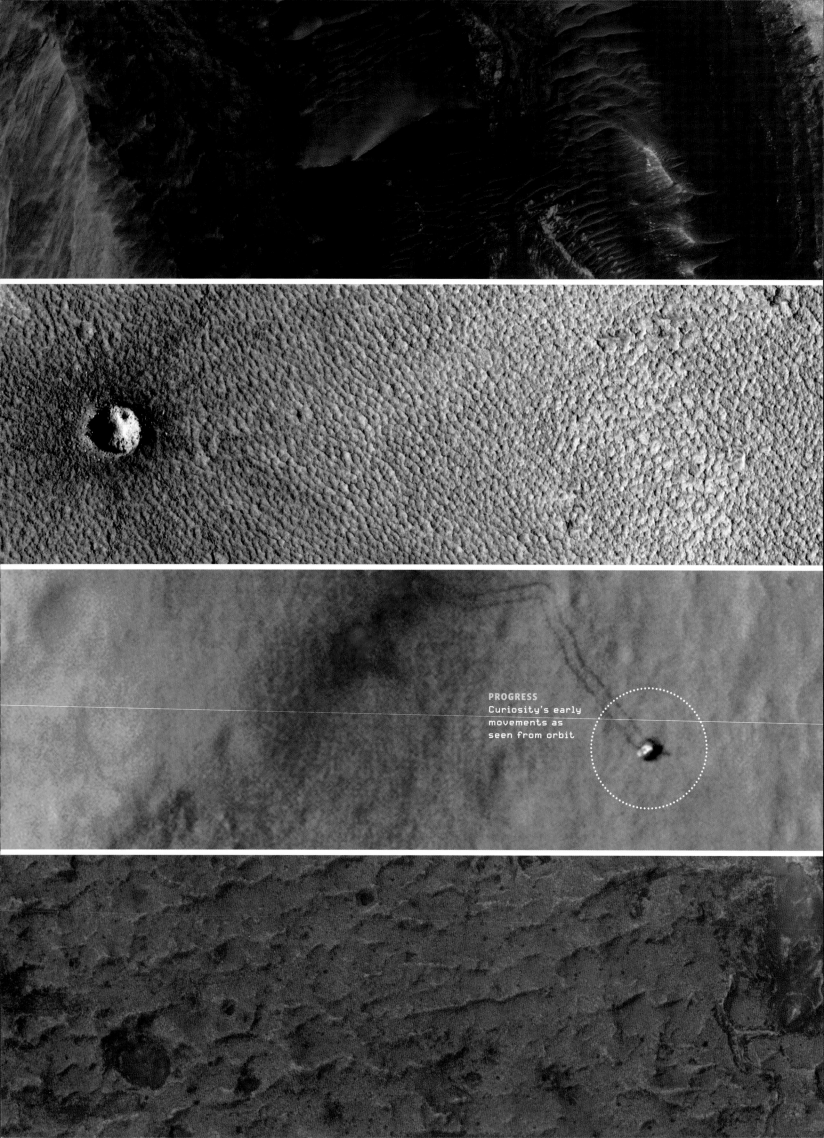

PROGRESS
Curiosity's early movements as seen from orbit

For a mission focused on the question of whether Mars could once have supported life, that first reaction said a lot. Gale Crater may be devoid of recognizable life today, but its geology seemed to reflect a potentially different past in the 4.6-billion-year history of the planet.

Most strikingly, Curiosity had apparently landed at the tail end of what looked from above to be the fossil remains of an alluvial fan, the unmistakable end of a stream or river that once spilled down the crater wall and into Gale. This was to some extent observed by cameras on an orbiting satellite, but the rover cameras were about to take their findings to a new level.

"Follow the water" had been NASA's mantra when it came to Mars for almost two decades. Their logic: Without H_2O—the best solvent by far for enabling the chemical reactions we call life—the rise of life was deemed to be impossible, or at least highly improbable. With water clearly present, the odds could rise significantly.

Once before, NASA scientists had been convinced they were about to find evidence of substantial water deposits and flow. When the previous rover, Spirit, landed in Gusev Crater in 2004, many were convinced they were about to begin exploring an ancient lake bed. That smooth, flat landing site, however, was ultimately found to be the result of a lava flow. Mars has a history of being cruel like that.

The two previous rovers did ultimately find evidence of the past presence of water, and the Phoenix lander found buried water ice in the northern plains. But Curiosity scientists were primed to finally ground-truth the long and widely held belief that water once flowed and pooled on the surface of Mars.

So at first glance, the landing site looked somewhat earthly, or at least like a place that long ago was much more like Earth. But the world Curiosity had just entered was in other ways different in the extreme. It took the rover some time to deploy its instruments and start taking measurements, but by extrapolating back and using other data collected earlier, scientists can pretty much describe the basic environment that the rover entered at touchdown. It was not a welcoming place.

CURIOSITY'S NEW HOME

If Curiosity was describing its new home, this is what we would hear:

First and foremost, Gale Crater is cold, and at night very cold. The late-winter air temperature at touchdown was at or just above the freezing point, but within hours the temperatures would fall to -115°F (-82°C) or below. This swing of more than 150 degrees a day continues through the year, and goes up or down to a more limited extent with the seasons. The crater is located near the Martian equator, where temperatures are as warm as they get. In contrast, nighttime temperatures at the poles in winter can reach -200°F (-128°C).

Mars is so cold in part because it's on average 149 million miles (240 million kilometers) farther from the warming sun than Earth, but also because the planet has such a thin atmosphere. Carbon dioxide—that abundant greenhouse gas on Earth—is by far the most common compound in the Martian atmosphere. But CO_2 needs water present to hold in arriving warmth, and the planet and all that surrounds it is now bone-dry.

Given the frigid nights, Curiosity would soon freeze up if the rover didn't have a system for generating heat.

The Mars atmosphere is also only one percent as "thick" as ours. In most basic terms, that means on average Mars now has but one chemical molecule floating in its atmosphere for every 100 molecules surrounding (and protecting) Earth. Absent any sort of blanket of molecules, the sun's heat quickly sails back into the atmosphere. Given the frigid nights, the inner workings of Curiosity would soon freeze up if the rover didn't have a system for constantly generating and distributing precious heat. Small nuclear generators produce the heat, which is stored as electricity in batteries and sent through pipes to keep the rover and its instruments warm. The nuclear generator is heavily protected and has a projected lifetime of 14 years.

The ground temperature also swings dramatically during a Mars day (officially called a sol), and can get colder than the air by a considerable amount. But even before touching down at Gale, Curiosity scientists knew that the landing ellipse was divided into sections that had quite different ground temperatures. One region in particular held the daytime warmth considerably longer than nearby bedrock; in geological terms it had a "high thermal inertia."

On Earth, scientists know, bedrock with that ability to hold heat longer than nearby formations often is made from harder rocks, sometimes the result of water-induced chemical reactions that "cement" sand grains together. This is the kind of clue the Curiosity team had followed as it worked to piece together the history of the crater and the planet—a kind of *CSI: Mars* where every bit of newly discovered information can lead to a much larger discovery.

Those large temperature swings in a thin atmosphere create another distinctly Martian dynamic—similarly large swings in air pressure. On Earth, the daily swing in pressure is one tenth of one percent of the global mean; Curiosity has found regular daily pressure changes of more than 10 percent on Mars. That kind of pressure change occurs on Earth only with the coming of a huge hurricane, or if you fly from low-lying Los Angeles to the mountains in Colorado.

Gale Crater has wind as well, especially strong when coming off three-mile-high Mount Sharp. On Earth, wind is generally a function of surface-temperature differences, as between land and an ocean. On Mars, it's often a result of the same dynamic, but in a different form since there are no oceans. Much of the Martian surface shows low thermal inertia, meaning that it heats up and cools off quickly. Morning and afternoon winds—which can be seen in the crater blowing the now grounded parachute that helped the spacecraft brake during its descent—are created by the abrupt changes.

And at times the winds can get substantial, as seen in the eroded mountain and bedrock, and also in some of the rocks themselves. Rocks that have been vigorously sculpted by wind have a particular set of shapes and textures, and geologists see them in the images taken by Curiosity. They're called "ventifacts," artifacts sculpted by wind.

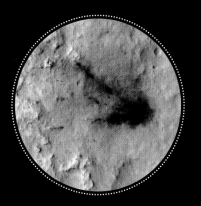

EXPLORING A NEW LAND >> Buttes and mesas on lower Mount Sharp (below) made Gale Crater attractive to Mars scientists. The minerals they hold, the size and texture of the rock grains, and the order of the layers may reveal Mars's climate and environmental history. For perspective, the pointed mound in the center is about 325 feet (100 meters) tall. Marks from landing now scar the landscape (above, left) of the planet Mars, more traditionally envisioned as the red planet (above, center) in an image taken by the Hubble Space Telescope. Once alighted, Curiosity readied its instruments, such as the turret at the end of a robotic arm (above, right).

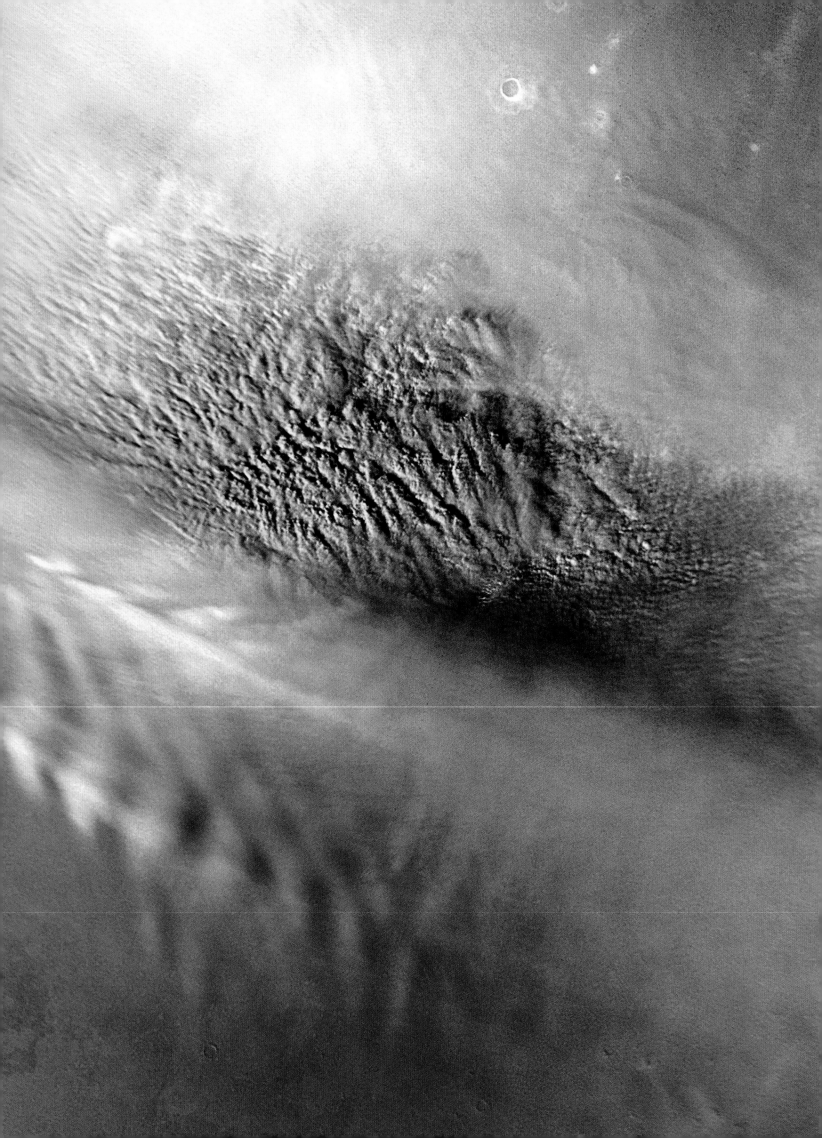

<< Dust storms are frequent on Mars, and this close-up image of a large one was taken by the Mars Color Imager on the Mars Reconnaissance Orbiter. This storm lasted but one day, but others go on for weeks. Also visible are the planet's northern polar cap, at top, and gravity-wave ice clouds coming off a crater, at bottom.

The thin atmosphere of Mars, especially near the surface, is bone-dry—holding some 10,000 times less moisture than Earth's. Assuming that Mars was once much wetter and warmer, that means H_2O has been escaping into space for eons, or being turned into subterranean ice. Understanding the fate of Martian water is one of the goals of both Curiosity and future NASA missions, and there are many leads to follow.

Many scientists began the mission convinced that the fans and deltas identified at Gale Crater and elsewhere across Mars were cut by once flowing water and hoped Curiosity would confirm it. But water can also be found baked into some of the clays and other minerals detected in Gale Crater by satellite. In fact, the apparent abundance of water-related minerals was one of the reasons Gale was selected as the landing site.

Throughout history and in culture, Mars has always been most famously red. As a result, the planet is often imagined as a bloody, hostile, and pretty violent place: Mars, after all, was the Roman god of war. The reddish color comes from iron oxide (rust) that fills the sky and coats the surface. The compound is caustic and corrosive and has defined the planet—or at least its surface—for eons.

And then there's radiation. During the long ride from Cape Canaveral to Gale Crater, the spacecraft carrying Curiosity was constantly bombarded by particles found throughout space—cosmic and solar radiation. These high-energy atomic nuclei are radioactive isotopes like the radiation released during the nuclear accidents at Fukushima and Chernobyl, or during testing of nuclear bombs. Some cosmic and solar radiation does make it to the surface of Earth, but our thick atmosphere and the magnetic fields that surround the planet reduce their impact substantially. With its much thinner atmosphere and no magnetic field protection, the Martian surface is constantly scorched by radiation.

NASA has long experience now with hardening spaceships, rovers, and landers against radiation, which can corrupt electronics without proper protection. But actual radiation levels on the surface of Mars had never been measured. One of the rover instruments is, for the first time, measuring that surface radiation with an eye toward future human travel to Mars.

SENSING MARS

As a stand-in for those humans, Curiosity was designed to fend off all the extremes that Mars now throws at it. Astronauts will someday have to be even more highly protected, and some doubt the worth of that. But imagine what some of their additional senses might be able to take in.

What, for instance, does Gale Crater smell like? There's a lot of iron in the air and on the ground, but there's also sulfur and even hydrogen sulfide, which gives off the rotten-egg smell. No instrument has ever measured the smell of Mars, so we don't know for sure.

Sound has also not been explored on Mars, although at one least noise-capturing instrument was sent to the planet on a different mission. The spacecraft carrying it, however, crashed on landing, so the instrument was lost. Nonetheless, experiments on Earth using Mars-like environments have shown that noise travels differently on the planet.

The Martian site where Curiosity first came
to rest was named in honor of the author of
the iconic Martian Chronicles, Ray Bradbury.

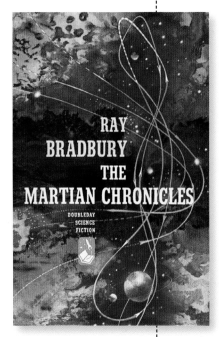

The Curiosity landing site was named Bradbury Rise after author Ray Bradbury, a great annalist of Mars, known especially for The Martian Chronicles. Bradbury died on June 5, 2012, just months before Curiosity landed on Mars.

With an atmosphere that's much less dense than Earth's, machinery that would sound noisy here—some of the gears and other moving parts on Curiosity, for instance—would sound far less noisy on Mars, generating fewer decibels of sound. In addition, the thin atmosphere carries low-frequency sounds much better than high-frequency sounds. So if there were streams somehow flowing on Gale Crater today, the sound of the rush or roar would carry much louder and farther than the higher-pitched burbles or trickles.

But very low frequency infrasounds—the distant rumbles and crashes that on Earth can be picked up by elephants and some other creatures, but not by humans—would be more pronounced on Mars. So if a person or sound instrument was present at Gale Crater, they could potentially hear the distant landing of a meteorite or the kicking up of a faraway dust storm. Otherwise, about the only sound would be the wind.

The Martian site where Curiosity first came to rest was, soon after landing, named in honor of the just deceased author of the iconic *Martian Chronicles,* Ray Bradbury. (NASA and other organizations avoid the naming strictures of the International Astronomical Union if any site, crater, ridge, or rock under consideration measures less than 100 meters in its longest dimension.)

The Curiosity science team chose the name, and the selection was virtually unanimous: Many of the scientists had been inspired by the writer's imaginings about life on Mars. Bradbury's story looks back at a Mars that once had a sophisticated indigenous people who were, over time, colonized by waves of Earthlings.

Sometimes intentionally and sometimes unintentionally, the people from Earth largely destroyed the Martians and their civilization. Then the planet was abandoned by the newer inhabitants because of troubles back home on Earth. The "Chronicles" are about a Mars that, if visited after the narrative finished, would be pretty desolate—but with many clues left behind about the life that once existed.

Fittingly, the primary job of Curiosity is to dig into and understand the early days of the real Mars, to find clues about what the planet was like before it became the cold, harsh, reddish, inhospitable place it is now.

By better understanding the Mars of today using Curiosity and its powerful instruments, scientists hope to find and write a new narrative about when the planet was younger, wetter, warmer, and almost certainly not red.

FROM MANY PERSPECTIVES >> The Mars Hand Lens Imager (MAHLI) reached under Curiosity to photograph the rover's three right wheels, with Mount Sharp in the distance (above). The spaceship landed in what had been identified as Quad 51 of the ellipse (below) for purposes of putting together a geological map. Another camera, the navigation camera (Navcam), perches at the top of the rover's mast and can look down at Curiosity below (right).

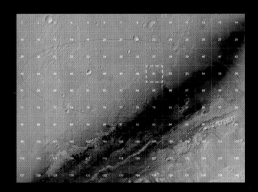

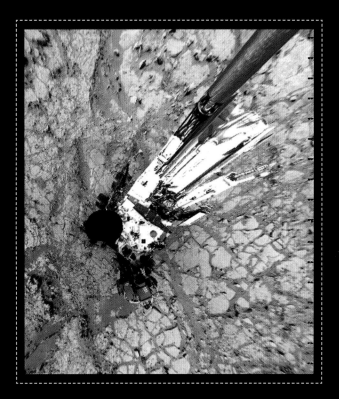

"I broke down and cried when I saw Mount Sharp, because I had spent so much time looking at it from orbit."

The rise and evolution of microbes on early Earth is endlessly fascinating to Sumner, and she has spent years working to understand them. The transition to thinking about and later helping to explore for microbes that may have once lived on Mars—and possibly still do—was a natural and easy one. Her tasks for the MSL/Curiosity mission have been many, ranging from day-to-day operations to helping to select the landing site, from working as a member of the Preservation of the Martian Organic and Environmental Record Working Group to conducting Mars science. For the 2013 Lunar and Planetary Science Conference, her name was on abstracts for 13 papers. She temporarily left the Curiosity team in fall 2013 for research in Antarctica.

Dawn Sumner, University of California, Davis

Sumner, a long-term planner for the Curiosity mission, also led the team effort to examine detailed orbital photos of the Gale landing ellipse and map its geological features (below).

137°15'E
4°30'S

AF

AEOLIS PALUS

GEOLOGY OF
GALE CRATER

137°30'E

Cs

BF

HP

HP   Yellowknife Bay
Bradbury Landing Site

Cs
Cs
RT

Cs
HP   Cs

RT

AEOLIS MONS
[MOUNT SHARP]

Cs

RT
Cs   Mt. Sharp Entry Point
(Murray Buttes)

4°45'S

KILOMETERS
0 1 2 3 4
0   1   2   3
MILES

MAP KEY
— Anticipated route
GEOLOGIC UNITS
☐ Alluvial fan (**AF**)
☐ Bedded fractured (**BF**)
☐ Cratered surface (**Cs**)
☐ Hummocky plains (**HP**)
☐ Rugged (**RT**)
☐ Striated (**SR**)

MAPPING MARS FROM AFAR AND UP CLOSE

Even before Curiosity arrived at Gale Crater, a map was ready to guide it. But this wasn't a travel map: It was a geological map.

The now elaborate and detailed rendering of much of Gale Crater was the inspiration of Curiosity project scientist John Grotzinger, and it was not unlike the maps that geologists draw when they go into the field. They look for distinguishing characteristics in what might at first seem to be a uniform landscape, mapping what looks different or unexpected. From those small beginnings, many geological discoveries have been born.

Initially, the eyes collecting all the geological information on Mars were those of the High Resolution Imaging Science Experiment (HiRISE) camera on the Mars Reconnaissance Orbiter, the most powerful camera NASA has ever sent to Mars. The MRO orbits now between 160 and 200 miles above Mars and, at its closest, can spot and take images of features as small as a dinner table. Previous Mars orbiter cameras could see objects no smaller than a school bus.

After Gale was selected as the landing site, an already extensive digital rendering of the crater was expanded greatly and a huge collection of orbital images of the ground was created. A geology professor at the California Institute of Technology with decades of experience in the field, Grotzinger saw an opportunity to use the HiRISE images to understand some of the intriguing features of the terrain before arrival at Mars. He began planning the most detailed orbital geological mapping of a planet other than Earth.

At first glance, many of the images look undistinguished and undistinguishable. That many are two-dimensional doesn't help a viewer understand what the three-dimensional landscape might look like. But a team, the Gale Mapping Working Group, was put together to start digging into the images, and the project took off.

The mapping effort brought in 39 scientist volunteers and was led by University of California, Davis, geologist Dawn Sumner. A specialist in the geology of the early Earth when life began, she was drawn to the prospect of mapping out in fine detail a terrain that was as old—or older—than anything she could study on Earth. She had also played a central, six-year role in selecting the Curiosity landing site, and so was already familiar with Gale's craters and canyons, its dunes and its dried riverbeds. And, of course, its mound.

The mapping team labored while the spacecraft flew to Mars, each person studying his or her digital quad for hours on end. Inevitably, the process of making order out of the seeming chaos of the images was both challenging and tedious. Sumner, a veteran of searching out clues in harsh environments like Antarctica and Western Australia, explained: "Especially looking down on Mars, everything is a little bit of a blank slate and can look the same. But the idea is to take what you see and break it into useful differences.

"It could be texture, contrast, how the rock weathers, colors, fractures—whatever at the time seems meaningful, even if you don't know why. It's really hard. But when you have a clear idea that something is significantly different, you put a line on the map and draw a boundary.

The first rock analyzed by Curiosity instruments was pyramid-shaped Jake M, named for a recently deceased JPL engineer, Jake Matijevic. The unusual specimen, unlike any other Martian igneous rock ever found, was a mugearite—the kind of volcanic material found on islands such as Hawaii.

TOPOGRAPHIC
DICHOTOMY

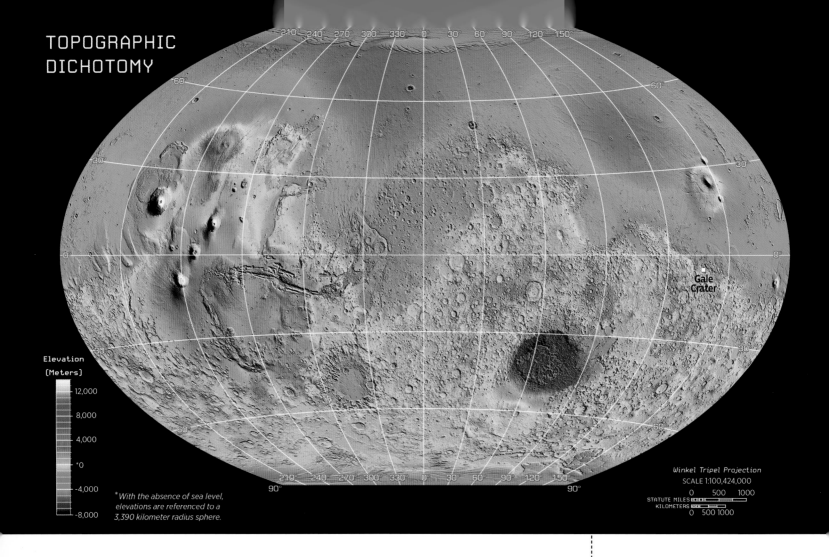

**Elevation
(Meters)**

- 12,000

- 8,000

- 4,000

- *0

- -4,000

- -8,000

*With the absence of sea level,
elevations are referenced to a
3,390 kilometer radius sphere.

Gale
Crater

Winkel Tripel Projection
SCALE 1:100,424,000

0   500   1000
STATUTE MILES
KILOMETERS
0   500 1000

"As you make more lines, you gradually start to see things that are systematic, connected. With those in mind, you see differences you missed before. In the field, you often go back to where you started mapping and see a million times more than when you started."

A week before landing, the individual quad maps, with lines and boundaries and highlighted features, were digitally stitched together and made consistent, creating the first interplanetary geological map of its kind.

A primary goal of the mapping team was to identify "units"—sections of rock that had textures or surfaces or fissures or heat-conserving features that connected them. At touchdown, the team had identified six of these units that appeared to make up the crater floor at the landing ellipse.

The mapping also helped tease out features initially hidden or obscured. Long before the rover touched down, Curiosity scientists knew that what appeared to be the broad fan of an ancient stream or river ran for 15 miles from the deep Peace Vallis canyon, which seemed to be carved by water on the north side of the crater rim. The flowing water then spread out into what scientists call an alluvial fan that, initially, did not appear to go as far as the Bradbury landing site, stopping instead a mile or two away.

THE ALLURE OF THE ALLUVIAL FAN

That's how it first appeared: The texture and shaping of the rock bed changed, suggesting that the rock unit and the forces that made it were different, but the mapping showed that some of

Gale Crater sits near what is called the dichotomy line—the division between Mars's northern and southern hemispheres. While the north is low and flat, the south has greater elevations and much more contour. Mars scientists generally agree that the south retains features from earliest Mars, while the north was later covered by lava, mud flows, or water, which wiped out the ancient record.

As the contours and size of the fan were being mapped,
previously invisible features came into view.

the characteristics associated with the fan showed up farther on. Then, using some 3-D images, mappers found that a central convex bulge rise characteristic of alluvial fans also continued farther than initially observed.

Those insights helped Grotzinger and the Curiosity scientists decide that their first destination should not be Mount Sharp, as initially planned, but rather an area of three rock units that came together where, some argued, the alluvial fan finally ended.

As the contours and size of the 700-square-mile (about 1,800-square-kilometer) fan were being mapped and analyzed, previously invisible features came into view. For instance, specialists in river formations found a number of "inverted channels"—ridges of hardened sediment or minerals within ancient streambeds that become visible after the water is gone and the wind etches them out.

The size, shape, and makeup of the inversion can give scientists important insights into the strength and duration of the ancient river flow. A bone-dry heap of rock within a bed or gully might not seem like a great find to the uninitiated, but to those who knew, it was a flashing light. Where there are inverted channels, there once was a potentially life-enabling supply of water, and in this case the river or stream was a full 15 miles long before it began to fan out.

Then there were the outcrops—exposed rock faces that even from orbit could be seen to have layers in them. To geologists, layers are essential keys to understanding history. Laid down one on top of another, they tell of different conditions, different dynamics, and the order in which they probably formed.

In geological terms, the assumption is generally that the top layer is the youngest and the bottom layer the oldest—especially on Mars, where there are no convincing signs that plate tectonics have jumbled up the landscape. If the quad you were mapping had formations that were clearly layered outcrops, you had something important to work with.

The mapping was such a hit that many in the working group continued with it after landing, moving into adjoining areas and adding novel new material. A graduate student of Grotzinger's, Katie Stack, took up the challenge of collecting the HiRISE data for the full six-mile (ten-kilometer) journey to Mount Sharp and began mapping the rock units, the potential outcrops, the distinguishing and different features on the path. She presented her findings and recommendations to the full science team.

The maps also offered hints of potentially important features ranging from ghost craters (which look like flat land but are actually filled with fine dirt) to structures called pingos. Well known in Arctic and subarctic regions on Earth, pingos are circular mounds formed by the welling up of water in permafrost regions. They have a series of well-known characteristics and are always

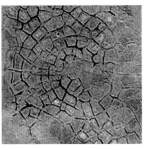

MARTIAN PINGOS? On Earth (left), pingos form when groundwater pushes up mounds of ice in areas with permafrost. Some believe the remains of pingos have been found on Mars (right).

associated with fractures in the shape of polygons, cookie-cutter breaks in the ground also common in the far north of both Earth and Mars. If the shapes are determined to indeed be ancient pingos, that would say a lot about how water behaved in ancient Gale.

The map was precise and deep with information at touchdown, and has continued to grow and become richer with time. Once Curiosity began supplying images—especially the ones focused on the ground and terrain just ahead, taken by the workaday hazard and navigation cameras—that information was added to the increasingly rich 3-D map.

FILLING IN THE DETAILS

As the instruments on Curiosity began to gather chemical and mineral information, they too became part of the map. The main contributor was the ChemCam, which can analyze compounds and elements in the rocks and soil from afar using a laser zap and a spectrometer to read the released gases. If it found a lot of calcium, sulfur, or silicates in an area, that information would be added to the map, gradually creating a rendering that was geochemical as well as geological.

Fred Calef is a planetary geologist at JPL and one of the architects of the mapping system. A self-described "wannabe poet," he nonetheless took the job of map keeper—with all the computerized cartography, basic geology, and cat herding of scientists that it requires. The mapping group, he says, could spend an hour deciding where to draw a particular boundary line, or whether to connect two identified rock units or keep them separate.

"Think of printing a map like this: You put it on paper and it's done. The power of our system is that we can reinterpret, can move lines, and delete as new information comes in.

"We knew we would get just deluged by all the data, but we were determined to keep it all in a usable format. Often that's done years down the road; our goal was to try to use the data immediately, to put it in a format our scientists could use right now."

This panoramic view of Gale Crater, with Mount Sharp in the distance, is made up of a series of stitched-together images taken by Mastcam on Sol 170, the 170th Mars day of Curiosity's expedition.

>> Rover drivers used animations such as this one to validate its systems and capabilities as they plotted out an early Curiosity test drive.

So while images of Gale Crater (minus Mount Sharp and the crater rims) may have looked like a barren and mostly featureless place to the uninitiated eye, the science team knew before landing that it was filled with tantalizing promises.

Given her months studying the surface of Gale, mapmaker Sumner was the natural choice to be the science team's long-term planner at landing, with the task of quickly absorbing the sights and measurements coming in and recommending paths forward. Sumner is known for her calm under pressure, but even she lost it when the first Curiosity photo of Mount Sharp appeared on the big Mission Control monitors.

"I broke down and cried when I saw Mount Sharp because I had spent so much time looking at it from orbit." To actually see it from the ground—looking at it bottom up rather than top down via satellite cameras was just too wonderful to bear.

IMAGING MARS

THE WAR PLANET Starting with the Greeks and Romans, the planet Mars was associated with war. Seventeenth-century Flemish artist Bartholomaeus Spranger pictured Mars (left) prepared for battle. Beginning in the 19th century, Mars was also seen as a planet inhabited by industrious creatures. Percival Lowell (above), in his observatory, was part of that movement.

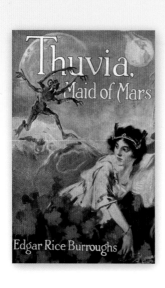

WILDEST DREAMS Edgar Rice Burroughs's series of Mars science fiction books, including Thuvia, Maid of Mars, often featured kidnapped damsels in distress and conflict between warring Mars factions. The movie Flash Gordon's Trip to Mars featured the villain Ming the Merciless, innocent Martians turned to clay, and heroics by the planet-hopping protagonist.

MAPPING MARS Nineteenth-century Italian astronomer Giovanni Schiaparelli first proposed the existence of canals, or channels, on Mars, an idea taken up by Percival Lowell. This 1894 rendering of Mars by Eugène Antoniadi, redrawn by Lowell Hess, reflects the canal influence. Antoniadi later rejected the canal theory, and his maps were used in the first naming of features on Mars.

Throughout the centuries, humans have assumed there was life on Mars. That life might consist of warlike, menacing, violent creatures, the natural inhabitants of a blood red planet. Or it might consist of industrious, beautiful, and even advanced and telepathic beings. But whatever their dispositions, they have always been there.

WAR OF THE WORLDS A high point in the presentation of Mars as a hostile, aggressive place was reached when American actor Orson Welles (above) famously re-created the H. G. Wells novel The War of the Worlds (illustration, left) on radio in 1938. Many who tuned in late believed the broadcast was a real news report and responded as if the United States was really being invaded by Martian monsters.

GETTING STARTED

CHAPTER 3

Ensuring that all systems are operating and communicating as designed

THINK OF THE LANDED Curiosity as a chrysalis. Its shell was hard, its vital parts all protected inside.

The tucked-tight rover had ten science instruments and 17 cameras. It contained a nuclear-powered battery and a seven-foot arm complete with a drill and soil sifter, folded up even tighter. On board were two redundant main computers and scores more to control processes of all sorts, as well as hundreds of miles of connecting wiring and two lasers. Its communication equipment could transmit to three orbiting satellites and to Earth directly through the big dishes of the Deep Space Network.

But at landing all this and more was in dozing mode, waiting to be awakened. What was operating that first day on Mars were the system-directing computer, the Hazcam cameras, the battery, and some of the basic communications equipment.

Yet Mission Control at JPL was already well into its much practiced rollout for bringing Curiosity back to wakefulness. Landing a one-ton rover on Mars is obviously dramatic and hard; bringing all its systems and instruments up in the correct order and properly in relation to each other may not be dramatic, but it certainly is challenging.

And Curiosity was orders of magnitude more complex and consequently more difficult to get going than anything that had landed on Mars before. The JPL manager

Officially, they were all off Pacific time and on Mars time.

most directly in charge of this surface start-up was Jennifer Trosper, a veteran of the Pathfinder, Opportunity, and Spirit rover missions to Mars and (talk about difficulty factors) a mother of three young children, including a baby born four months before landing.

As an insight into the increased complexity of Curiosity, she says that Sojourner, which landed on Mars in 1997, had five computer sequences preprogrammed for landing; Opportunity and Spirit each had about 50, and Curiosity had roughly 500.

Because their planned missions were so short (Sojourner's was 7 days and Opportunity's and Spirit's were 90), getting a quick start after touchdown had been considered essential. The Curiosity mission, however, is expected to run about 23 Earth months (and perhaps much longer), so the need to get moving fast is greatly reduced. In the end, the rover didn't move at all until Sol 16, and didn't have a real drive until Sol 21.

WAKING UP CURIOSITY

In the interim, Trosper and her team worked with the scientists in charge of the ten instruments to determine if they were healthy. The process began with a ping sent from Earth: If all was well, the instrument would answer. The team in charge of the Sample Analysis at Mars, or SAM, programmed the instrument's computer to respond to that first wake-up call with an iconic introduction from Dr. Seuss: "I am SAM, SAM I am."

Trosper also oversaw a process that switched the computer software from the cruise program to the surface program. This was the first switchover of its kind, made necessary by the vastly increased number of tasks that had to be performed by the spaceship/vehicle. (The computer had some 35,000 parameters or knobs to tune the larger system, and all of them had to be switched out.)

It took a collection of about 50 specialist engineers working around the clock to make it all happen. They were the new team on the front line now—filling the seats and commanding the monitors in the mission support control room.

Officially, they were all off Pacific time and on Mars time, where each day is 24 hours and 37 minutes long. This new time frame would last for 90 days, and—along with the excitement, stress, and long hours of the early period—it would keep most of the 500 engineers and scientists at JPL in a near-permanent state of jet lag. The concept of a Mars sol, as opposed to an Earth day, became part of the JPL language, with references to "yestersol," "holisol," and "soltime."

The engineers' job was organized around a schedule of receiving a data downlink (usually via the Mars orbiters), followed by 16 hours of high-speed analysis by instrument

Curiosity's Sample Analysis at Mars (SAM) instrument is a sophisticated portable chemistry lab. A twin SAM test bed (below) resides at Goddard Space Flight Center in Greenbelt, Maryland.

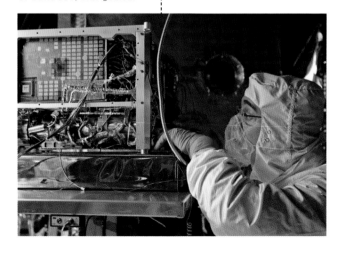

science and specialty science teams (geology, atmospherics, chemistry) as well as engineering teams to determine what was safe and possible. Then came planning for the next sol and horse-trading between engineers and science teams to determine what would make it into the sol's program and what would not. With little wiggle room, the new commands for the rover's upcoming sol had to uplink before those 16 hours had passed.

On a normal day, Trosper and many of her colleagues would work much of that 16-hour period and then, rather anticlimactically, go home while Curiosity actually conducted all the activities sent up via that sol's uplink.

Time at JPL became ever more confused for the Curiosity team as the sols went on—with the campus at times packed and busy through the night, then largely dark during other periods. The cause was the necessary marriage of ever changing Mars time with the unchanging

Using the MAHLI camera at the end of its seven-foot robotic arm, Curiosity engineers have commanded the rover to take a series of self-portraits that show just how alone it is on Mars.

The process went remarkably smoothly, but all it would have taken was one glitch among the 10,000 things going right to usher in trouble.

16-hour sprint from downlink and the next uplink. The time for prime analysis, theorizing, and decision making would be during daytime one month, then in the middle of the night the next. That meant individual shifts would start and end all around the clock—and change by almost 40 minutes every day.

"It was those nights when my shift began at ten at night that I really felt it," Trosper says, speaking for many others. "But I was lucky to have had the experience of living through Mars time twice before with the other rovers. And, to tell the truth, nobody sleeps much when there's a baby in the house anyway."

Trosper's husband, F-16 pilot Lt. Col. Randy Trosper, volunteered to take care of the children so his wife could manage the mission. Big picture: It worked out well. Finer grain: There were times like the evening he called to say he was driving the van with the three kids strapped in, hoping they would go to sleep. Absent that, he told his wife, "I was going to lose my mind."

So it was with the rover start-up. The process went remarkably smoothly, but all it would have taken was one glitch among the 10,000 things going right to usher in trouble. An antenna deployed on Sol 1, for instance, was sufficiently off-kilter to make telemetry communications sketchy.

It turned out that a simple error setting the angle of the antenna—tested at JPL, but missed—had embedded itself in the core computer program. It was easily fixed. But Trosper— who, like her perfectionist colleagues, is embarrassed by errors—grits her teeth and says that "we should have caught that, but didn't." The result was that the start-up process was slowed by a day, as it would be by other glitches and considerations. The original plan was for the rover to begin moving by Sol 10, but the first short spin didn't happen until Sol 21.

By far the biggest challenge and accomplishment of those early days was that switchover from the cruise to the surface software. Trosper worked on that process for two years before landing, for the second of those years with a semipermanent team of ten working with her. The cruise software, called R-9, could keep the rover safe and allow limited, if essential, activities like generating power, providing basic communications, and processing the Hazcam imagining. But without the surface software —R-10—there would be no Curiosity trek, no use of the arm and drill, no running of the complex instruments. As one of Trosper's colleagues, Magdy Bareh, put it, the rover would be getting a brain transplant.

The switchover was planned to begin on Sol 4, and the new team in mission support waited for the results of their operation with the same anxiety—checking and double-checking—that had consumed the landing team. Unlike their predecessors, however, the switchover team labored in relative anonymity, as Trosper explained.

"If we can learn about the past of Mars, it could change our whole thinking about the solar system."

Over her desk at JPL, Trosper has kept a quote from an unlikely source of inspiration: former president Calvin Coolidge. The passage is about what is needed to achieve success, and his conclusion is that it's determination and persistence. That certainly was the case in recent years as Trosper, then Curiosity Surface Operations Mission Manager, prepared for the rover's first days on Mars. After landing, she oversaw the all-important initial checkouts and what she calls the "brain transplant" of the flight software for the surface computer software. Those were grueling days spent on "Mars time," but they later earned her the even more stressful job of Curiosity deputy project manager. Trained in aerospace engineering at MIT and the University of Southern California, she was also lead systems engineer for the Opportunity and Spirit rovers.

Jennifer Trosper, Jet Propulsion Laboratory, NASA

Now second in command for the Curiosity mission, Trosper oversees technical and engineering activities. In the mission's first weeks, she was in charge of Curiosity surface operations, including the first rover drives (below).

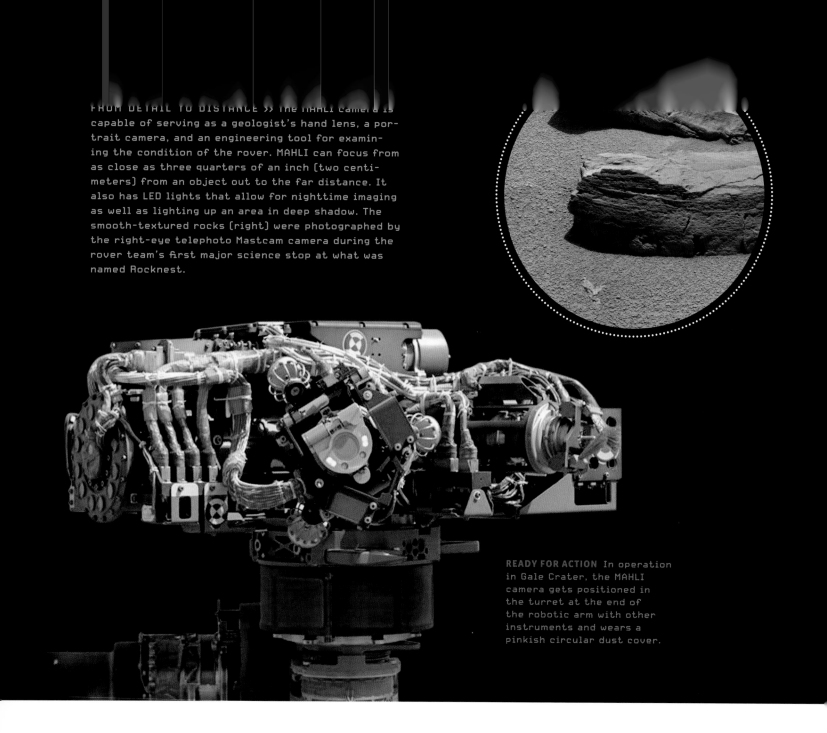

FROM DETAIL TO DISTANCE >> The MAHLI camera is capable of serving as a geologist's hand lens, a portrait camera, and an engineering tool for examining the condition of the rover. MAHLI can focus from as close as three quarters of an inch (two centimeters) from an object out to the far distance. It also has LED lights that allow for nighttime imaging as well as lighting up an area in deep shadow. The smooth-textured rocks (right) were photographed by the right-eye telephoto Mastcam camera during the rover team's first major science stop at what was named Rocknest.

READY FOR ACTION In operation in Gale Crater, the MAHLI camera gets positioned in the turret at the end of the robotic arm with other instruments and wears a pinkish circular dust cover.

"Except for our surface folks, there weren't many people in mission support those first few days. The others were ecstatic about the landing and exhausted by it all, and on one level the rest was gravy to some of them. Not that they didn't want it to succeed, but their part was over.

"But for us," she says, "this was the big moment we'd been working towards for so long. It was a make-or-break moment for the mission, but without any public drama."

The flight software was switched gradually over four days: a few "toe dips" with the systems before committing to each change.

"When it was completed, when the rover had the new software it needed to operate on the surface, I can tell you we were breathing huge sighs of relief," Trosper said. "We could then start testing mobility, the arm, and the higher-end aspects of the science instruments. Everything looked good, and I would have to say we were beyond ecstatic."

They were not alone. Scientists and engineers at the daily planning meetings often started their gatherings with a song or poem. David Martin, a systems lead for the SAM team

Expectations at JPL were great; breakthroughs
seemed within reach. Important mysteries
just might be solved.

from Goddard, began a mid-September meeting with these lyrics, put to a blues shuffle beat that he sang:

On planet Mars, the news from Mars Science Lab,

Rover arm and cameras, they're doing just fab.

The photos are clear, there's a past lake bed in our sight.

Find organics there, scientists say that we might.

You know, initial look-arounds are done,

We know where we first want to go.

Rover planners, it's time to shift,

Let's get the tranny out of low.

So NASA was safely on Mars with a supersize, highly sophisticated vehicle. All its instruments and systems appeared to be in good working order under the command of new surface software.

CURIOSITY'S GAME PLAN

The mission's primary goal was to determine whether the now blasted landscape of Gale Crater might have once been able to support life, or perhaps even once *did* support life. How exactly would they go about making that assessment?

The answer involves a mixture of very high-tech science and experienced eyes looking for telltale clues. The most solid clues lead to formal findings and papers, although their solidity can change dramatically based on new readings.

Each new finding, each new clue, leads to another round of debate and questioning: Why is that rock layer where it is, and what does that mean for the neighboring units? What elements are present in the rocks, in what proportions, and what isotopic forms (as determined by the number of neutrons in the atoms)? What do the rocks say about the long-ago atmosphere at Gale, and the possibility that it was once much "thicker"? Are specific minerals around that can only be formed in the presence of water? Are carbon-based compounds present that could have supplied building blocks for life?

Mars science is a jigsaw puzzle with more parts missing than found. But Curiosity had arrived much better equipped to search for those absent pieces than any of its predecessors.

Curiosity team members at JPL worked around the clock and on Mars time. A Mars solar day, or sol, is about 40 minutes longer than an Earth day, so each workday started and ended 40 minutes later than the day before and important tasks performed in the afternoon would, a few weeks later, take place before dawn.

What's more, many on the science team had spent a half-decade or more preparing for this opportunity of a lifetime. So expectations at JPL were great; breakthroughs seemed within reach. Important mysteries just might be solved.

A ROVING RESEARCH LAB

What may well most distinguish Curiosity from any previous planetary exploration is its ability to dig deeper and deeper into a promising or unusual find. It has a much broader range of tools to follow a path of inquiry from the general to the specific, from a precise geological assessment of an intriguing landscape to the chemical composition of a potentially revelatory rock in the formation.

The mission was designed that way, to allow the exploration of Gale Crater to be a broadbrush "mission of discovery." Specific goals were certainly set—to determine if Mars was ever habitable, to search for organic compounds, to reach and climb lower Mount Sharp—but they came with unusual leeway. The Curiosity scientists and engineers were given freedom to follow their findings and hunches, to set aside preplanned routes and timetables, and they ran with it.

The Curiosity discovery process began with pictures, but not from the 17 cameras on the rover.

Those came instead from the revolutionary Mars Global Surveyor Orbiter and its Mars Orbiter Camera (1997–2006), and more recently the Mars Reconnaissance Orbiter (MRO) and its HiRISE (High Resolution Imaging Science Experiment) camera (2006–today). Both were able to send back images with unprecedented detail of the Martian surface. The MRO also has a spectrometer called the Compact Reconnaissance Imaging Spectrometer for Mars (CRISM) that can read the chemical makeup of the Mars surface below. The MRO joined another satellite circling Mars, the Mars Odyssey Orbiter, and together they have dramatically changed scientific understandings of the planet.

It was images from the MRO, for instance, that helped persuade NASA to send Curiosity to Gale Crater. HiRISE identified unique layering on the mound, a possible alluvial fan in the crater, and a landing ellipse that was considered safe enough for a touchdown. Meanwhile, CRISM revealed mineral deposits in the crater that could be formed only in the presence of water. HiRISE also provided the images that were mapped by the Gale Mapping Working Group, and even took pictures of the rover and its discarded space capsule after they had landed. Well before arriving at Gale, NASA scientists had scientific leads galore.

Both the earlier Mars Orbiter Camera and a wide-range MRO Context Camera were designed and built by Michael Malin of Malin Space Science Systems, outside of San Diego. Malin, who built his company using money that came with his 1987 MacArthur Fellows Program "genius" grant, was also selected to develop and supply the two key cameras for the mast on Curiosity, one telephoto and one wide-angle. His team—made up of 20 top scientists who specialize in subjects such as mineral

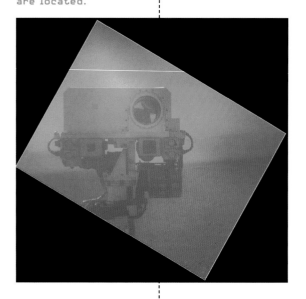

The rover photographed its own equipment as part of the early inspection process. The MAHLI's dust cover was closed when it took this Sol 32 photograph of the top of the rover mast, where the ChemCam instrument and two Mastcam cameras are located.

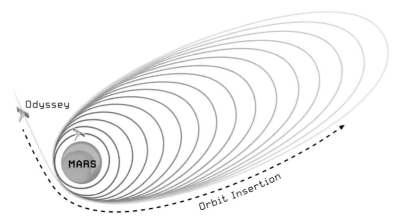

The Mars Odyssey is one of the three
satellites that were orbiting Mars
when Curiosity arrived. Placed into
orbit in 2001, it relays Curiosity data
back to Earth as well as conducting its
own science. Odyssey has orbited Mars
longer than any other spacecraft.

<< The Mars Reconnaissance Orbiter
arrived in 2005. It took six months of
"aero-braking" to turn its wide elliptical orbit into an orbit roughly circular and much closer to the planet.

identification, cold-weather sediments, frost, why landscapes look the way they do, and how best to visualize what the camera displays—pore over the images for clues that point to something unusual or especially meaningful, or that are part of a broader geological story.

The team also included filmmaker James Cameron, who worked with Malin on a zoom camera for Curiosity that would enable near-documentary camera work. The instrument, however, was scrubbed in the last year because it needed more testing.

Perhaps the most unusual camera on Curiosity is a high-precision, high-detailed version of the kind of hand lens that all geologists take to the field. Called the Mars Hand Lens Imager

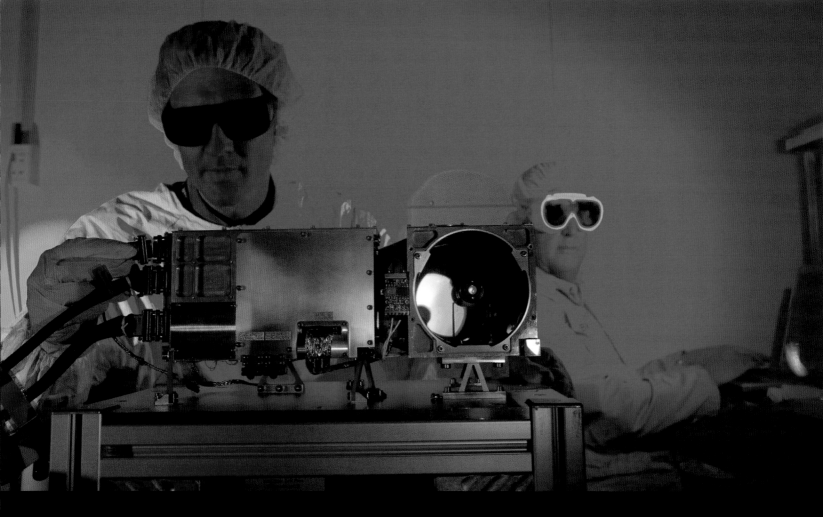

CAMERA FOR CHEMISTRY
A researcher at the Los Alamos National Laboratory in New Mexico prepares the mast unit for a test firing of the laser, used to complete the chemical analysis accomplished by Curiosity's Chemistry and Camera (ChemCam) instrument.

FIRST TARGET >> The ChemCam instrument uses a laser beam to vaporize a hole the size of a pin-head in rock targets (artist's rendering, above), then analyzes the resulting plasma to determine its chemical makeup. Designed and built by a team led by Roger Wiens at Los Alamos and the French space agency, it has a telescope, spectrometer, and camera that together perform the analysis. The first ChemCam target on Mars was a fist-size rock named Coronation, shown before the laser shot (upper right) and after (lower right).

The energy from the zapping turned atoms in the rocks into a glowing, gaseous plasma of charged particles.

(MAHLI), it can focus as close as two centimeters from a target and send back pictures of the finest of grains within rocks and sand, the kind that give geologists clues about what might have happened in the area in ancient times. Because MAHLI is mounted on the moveable arm of the rover, it has also been used to take pictures of the vehicle itself—including the iconic "self-portrait" that appears to have been taken by someone or something not connected with the rover.

The principal investigator for MAHLI is Ken Edgett, a specialist in Mars geology who works with Malin. Together they wrote a landmark paper for the journal *Science* in 2000 describing the layers of mineral and other rock near the base of Mount Sharp. These exposed layers, or striations, rival the Grand Canyon's in scale and are a major reason that Curiosity landed in Gale Crater. The discovery was based on both camera images and spectrometer readings taken by the Mars Global Surveyor.

So images taken both from afar and by Curiosity, as well as spectrographic readings of minerals and compounds from orbit, set the stage for the investigations to come.

CHEMCAM: FIRST LASERS ON MARS

A first look at the makeup of the rocks and soil on Mars came via an instrument with a laser zapper, a telescope, and a spectrometer all working together. Called ChemCam, it represented another first on Mars—in this case, the first time a laser was used on the planet's surface.

During testing, the ChemCam instrument shoots an invisible laser beam at an iron pyrite crystal in a sample chamber about ten feet away. Erupting from the pyrite is a ball of glowing plasma, the material to be analyzed for its chemical content.

On Sol 13, ChemCam took aim at a fist-size rock about nine feet away and blasted it with 30 laser pulses during a ten-second period. Each hit focused the energy of a million lightbulbs on an area the size of a pinhead. The energy from the zapping turned atoms in the rocks into a glowing, gaseous plasma of charged particles.

ChemCam then caught the light from that spark with a telescope and sent it to three spectrometers inside the rover.

Those spectrometers are able to identify the kind of rock targeted (volcanic or sedimentary, for instance), report what elements the rock is made of, recognize both ice and minerals with water molecules in their crystal structure, and even determine if the rock was changed through the activity of living organisms. ChemCam can hit targets up to eight meters away, and can do it dozens of times a day.

To understand just how revolutionary ChemCam is, consider what the previous rovers, Spirit and Opportunity, had to do to identify the chemical elements of a rock: The rover had to

The type of feldspar was something never before seen on Mars, so Wiens was concerned about getting it right.

select a target, approach the rock, brush it clean of dust, touch it directly with a spectrometer to make its readings, and then process the information. The whole sequence could easily take a week—and produce less information.

"If you can do something useful on Mars without going up to touch it, you can save a lot of money and time," said Roger Wiens, a geochemist at the Los Alamos National Laboratory in New Mexico. Wiens and his Los Alamos team, as well as colleagues at France's Centre National d'Études Spatiales who designed and built the laser and mini telescope, operate ChemCam and quickly proved the instrument can provide the remote readings they promised.

With ChemCam as a scout, the process of identifying interesting targets for more intensive study became easier. But the contact science still had to follow, generally first with the Alpha Particle X-Ray Spectrometer (APXS). It, too, reads the chemistry of rocks and soil, but it does so with added precision made possible by its proximity to the targets and the technology it uses.

Similar APXS instruments went to Mars on the Sojourner, Spirit, and Opportunity rovers, and they sent back readings supporting the theory that the planet was once far more wet and warm. The APXS on Opportunity identified salty conditions in bedrock that are strongly associated with a wet past, while Spirit's APXS discovered what was interpreted as the signature of an ancient hot spring or steam vent.

A far more sensitive and versatile APXS contributed by the Canadian space agency provides both deeper knowledge of the makeup of the rocks and soil and also some backup for the previously untried ChemCam. The APXS has to touch the target, or get very close to it, so it plays a somewhat different role from the ChemCam scout. But the Curiosity APXS can generate as much information about the elements of the rocks in 1 hour as the APXS on Spirit and Opportunity collected in 17 hours.

The symbiosis between ChemCam and APXS was quickly visible once the rover instruments could begin taking measurements. ChemCam repeatedly found concentrations of the elements sodium, silicon, potassium, and aluminum in rocks, which to mineral specialists means the presence of feldspar. On Earth it's a common mineral in American kitchens in the form of granite countertops. But the type of feldspar suggested by ChemCam findings was something never before seen on Mars, so Wiens was both excited by the finding and concerned about getting it right. The news that Gale Crater had lots of alkaline feldspar didn't come out until APXS had confirmed the finding, too.

SAMPLING ROCKS AND SOIL

But the two defining instruments on Curiosity are the Sample Analysis at Mars (I am SAM, SAM I am) and the Chemistry and

TIRE TRACKS The symbols "JPL" in Morse code were built into the rover wheel. The marks left behind are used to measure how far Curiosity has traveled.

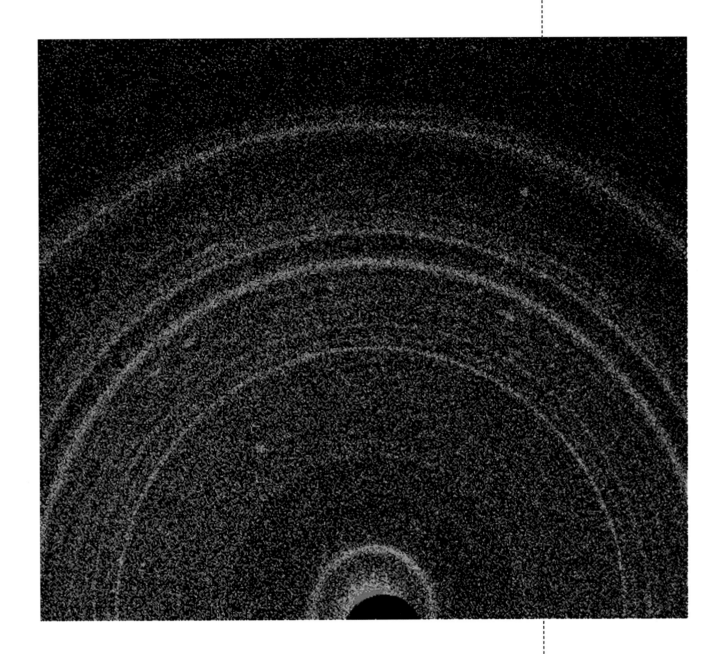

Mineralogy experiment (CheMin). Each is a miniaturized chemistry lab able to analyze Mars rocks and soil in entirely new ways and to push discoveries ever further.

For SAM, the arena is those all-important organic compounds that should be on Mars but have never been identified. CheMin specializes in analysis of minerals formed long ago in the presence of water. (It has a commercial use unrelated to NASA, too, in determining whether pharmaceuticals are pure.) SAM includes an oven that can heat samples to more than 1800°F (1000°C) while CheMin needs temperatures of -76°F (-60°C) to make its detections.

Both need Martian samples delivered to them, and samples from the surface have limited value. So Curiosity landed with the first drill ever on Mars, one that can dig two inches into rock and return a powdered sample. The depth may not seem great, but it is enough to reach down into the planet's ancient history. Two inches is also enough to find chemical and mineral samples little changed by the intense radiation that now transforms and defines the surface.

SAM, CheMin, and the drill are not only major advances on anything that came to Mars before Curiosity: Because of their substantial weight, they also created the need for a different

The chemistry and mineralogy instrument on Curiosity reads the atomic-level x-ray scattering of Martian soil samples and reports its measurements as graphics. Each mineral has a unique pattern of rings—its fingerprints—that can be identified, the colors representing the intensity of the x-rays, with red being the most intense.

kind of vehicle and landing. SAM alone accounts for more than half of the weight of the entire Curiosity science payload—88 pounds out of 165 (or 40 kilograms out of 75). CheMin adds another 22 pounds, or 10 kilograms. Both would take up a large part of a laboratory room on Earth.

SAM, CheMin, and the drill impose enormous time and power requirements on the mission. The process of setting up for a drill, gathering a sample, delivering it to the instruments, and then conducting the experiments takes weeks, and much of that time will be spent without moving the rover at all. One CheMin analysis alone takes ten hours; a SAM analysis can go for days.

Still, NASA determined the two were worth the many different costs involved because of the science and discoveries they are both capable of delivering and expected to deliver. Consider SAM and what it can bring to the decades-long search for organics on Mars. By heating up samples to the point that elements and compounds turn into gases, it can identify what's inside a Mars rock with unparalleled specificity. Those different component parts turn to gas at different tempera-tures, and the signature peaks captured by spectrum-reading instruments can tell researchers exactly what they are. The SAM oven's 1800°F capacity is almost 600 degrees hotter than anything

used there before. Since many organic compounds turn to gas at those high temperatures, SAM is a considerably more powerful organic finder.

The instrument also has a special capacity to test for the gas methane. Easy to produce on Earth and most often a by-product of life on our planet, methane was detected in the Martian atmosphere by NASA researchers in 2010, though not without some controversy. Potentially, SAM can both resolve that debate and shed some light on whether that possible Martian methane originated via biology or by geochemical processes.

And with CheMin, NASA would get the first detailed ground-truthing of orbital analyses that identify mineral deposits—most important, the deposits of minerals of clays, sulfates, and carbonates that can be formed only in sitting water. The results would bring some significant surprises.

Other onboard instruments have less expansive tasks and often take the exploration of Gale in different directions. The Radiation Assessment Detector (RAD) made the first ever measurements of radiation levels while flying to Mars in conditions similar to what an astronaut would experience, and the readings will continue on the surface until the instrument fails. The goal is to help guide future human travel to the planet.

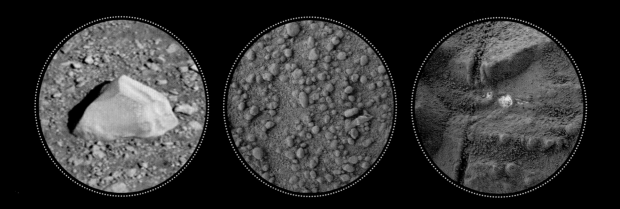

SOIL SAMPLING >> JPL has on hand soil surface specialists who can analyze the nature, texture, consistency, and general qualities of soil that the rover will be driving on: (above, left to right) sharp, wind-sculpted rocks; fine sand grains; and uneven bedrock with cracks. The rover went on its first test drive on Sol 15 (August 22) and traveled 19 feet (6 meters) (below). Later in the journey, the rover would sometimes travel as far as 475 feet (145 meters) a day.

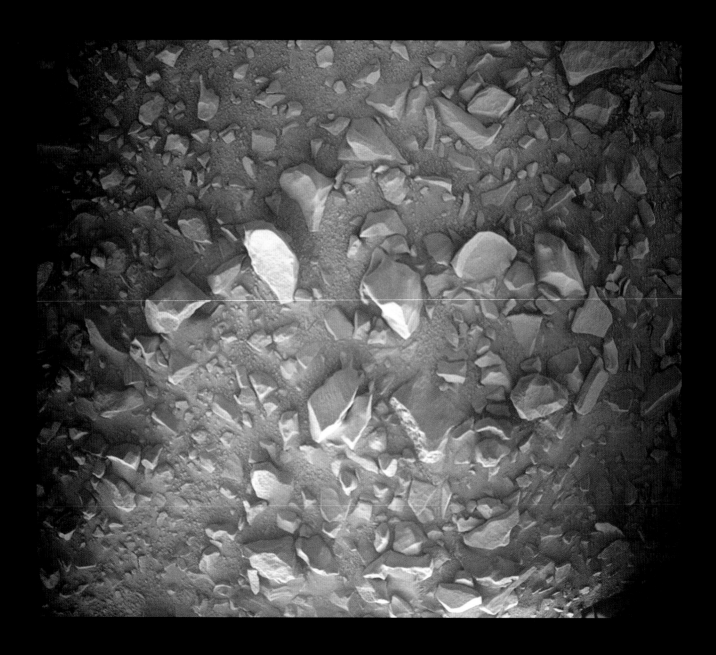

The Dynamic Albedo of Neutrons (DAN) investigation looks for signs of hydrogen (a stand-in for water) below the surface and gives geologists more information about the planet's watery past. And the Rover Environmental Monitoring Station (REMS) is an advanced weather station that is considerably more versatile and capable than previous Mars weather stations.

These instruments were planned and built in collaboration with other national space agencies as well—the Germans for RAD, the Russians for DAN, and the Spanish for REMS. So as Curiosity geared up for its exploration, scores of scientists from abroad gathered to join American colleagues at JPL as part of the science and operations team.

The early days were a heady time, as each rover system and then each instrument was started up and found to be in good working order. (One exception was a wind sensor on the REMS instrument that flew off during touchdown.)

But as would quickly become apparent, the greatly increased capabilities of Curiosity came at a price: Most everything would take longer than expected—sometimes a lot longer.

The chief engineer for Curiosity, JPL's Rob Manning, likened their creation to a huge 100-by-100-foot painting executed by a vast army of artists. Each artist painted with great precision and detail his or her allotted image, but when the whole picture was put back together, some of the juxtapositions were both baffling and jarring.

"Curiosity is successful and operable for sure," Manning said, "but it takes more commands to make it work than we imagined, and we have to think a lot more about hitting a button and sending a sequence of commands than we'd like.

"All the subtle and not-so-subtle interconnections that were unintended make it so you can never forget this is a prototype, and not a fully completed vehicle," he continued. "Put it this way: If you buy a car, it's nice to be able to listen to the radio and have the air conditioner on at the same time. Curiosity kind of suffers from limitations like that."

SOL 16: FIRST TEST DRIVE

Nonetheless, on Sol 16, Curiosity rover drivers had their first shot at test-driving their vehicle on Mars. They went about 16 feet (5 meters) forward, then they rotated 120 degrees and drove in reverse another 8 feet (2.5 meters), ending up close to where they had started.

The images of the tracks heading tentatively off into Gale Crater were greeted with applause and excitement, proof positive that the next phase of the mission had begun. But the images brought out another emotion, the mixed feelings that come when a parent sees a child go off to school for the first time.

And with each rotation of the rover's six wheels, Curiosity left behind a special marking in the Martian regolith, the loose soil atop the bedrock. Interspersed amid the angular tire tracks were sets of three slim rectangles with shapes imprinted inside them. In Morse code, the shapes spelled out three letters: JPL.

The message was never intended as a claim of ownership or sovereignty. But it did say clearly and forcefully that NASA's Jet Propulsion Lab and its rover had once passed that way on its Gale Crater exploration.

THE CURIOSITY ROBOT

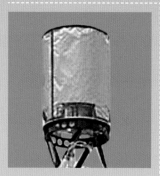

UHF ANTENNAS Most Curiosity information will be relayed to and from Earth via ultra-high frequency (UHF) links that use one of two redundant Electra-lite radios to talk to Mars orbiters passing overhead.

ROVER ENVIRONMENTAL MONITORING STATION
This suite of instruments measures temperature, wind, air pressure, humidity, ultraviolet radiation, and other indices to better understand daily and seasonal changes.

RADIOISOTOPE THERMO-ELECTRIC GENERATOR
The rover's power source, the generator uses plutonium-238 dioxide to provide heat that is turned into electricity by two thermocouplers, which then charge two lithium-ion batteries. The generator is designed to last for 14 years.

Curiosity hardware is a combination of brand-new and tried-and-true. The most novel and complex instruments are inside the rover.

CHEMISTRY AND CAMERA INSTRUMENT ChemCam uses laser beams to vaporize rock and then analyze the elements released. It is the first instrument on Mars to use a laser.

RADIATION ASSESSMENT DETECTOR (NOT PICTURED) Unlike past rovers and landers, Curiosity will monitor high-energy atomic and subatomic particles. The measurements will be critical in planning the human exploration of Mars some decades from now.

ALPHA PARTICLE X-RAY SPECTROMETER The APXS can read the chemical makeup of rocks and soil with greater precision than ChemCam, but it takes its measurements right at the target.

DRILL The first drill ever deployed on Mars, this instrument can dig 2.5 inches (6.4 centimeters) beneath the surface and deliver powdered sample material for study in the rover's two chemistry labs.

WHEELS All six wheels are fitted with driver motors, and the four corner wheels have steering motors as well. Each front and rear wheel can be independently steered.

THE RIVER

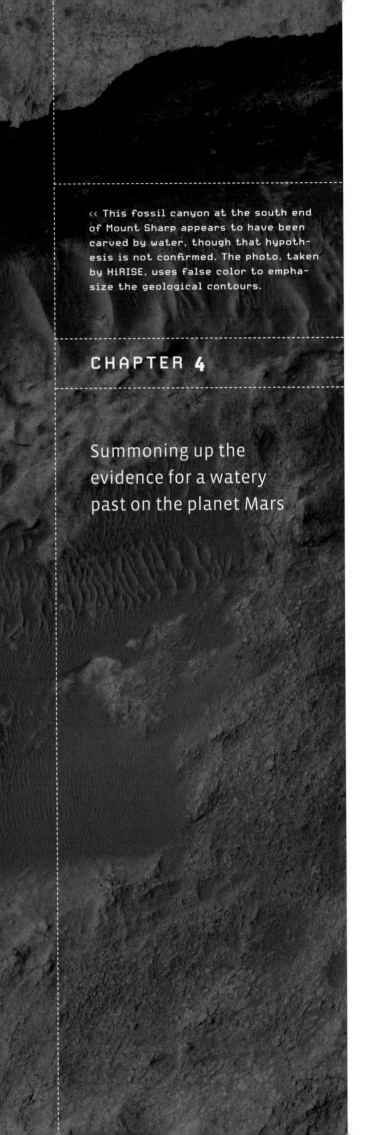

<< This fossil canyon at the south end of Mount Sharp appears to have been carved by water, though that hypothesis is not confirmed. The photo, taken by HiRISE, uses false color to emphasize the geological contours.

CHAPTER 4

Summoning up the evidence for a watery past on the planet Mars

BILL DIETRICH AND Becky Williams were two of the several hundred elated, exhausted, and expectant scientists crowded into a JPL control room on landing night. They had all celebrated and savored the successful touchdown, and now they were gathered to get a look at the first Curiosity images beamed back from Mars.

Nobody expected those initial photos to be more than low-tech, though surely glorious, proof that the rover had landed safely and was ready to work. They would come from both front and back Hazcams, low-to-the-ground cameras designed primarily to see proximate terrain for future driving purposes. The Hazcams are wide-angled and highly fish-eyed, and nobody imagined they would play much of a role in collecting scientific information.

But when the first black-and-white images went up on the dozens of monitors around the room, the hoots and hollers were accompanied by a stunned silence from Dietrich, Williams, and some of the other geologists—especially those who studied streams, rivers, and their deposits. What their well-trained eyes saw was so far beyond their wildest scientific dreams that the memory of it would still bring broad grins to their faces months later.

What did they see? A stretch of ground topped by a fair number of pebbles and some finer-grained gravel. Many taking in the images searched for (and ultimately found) the outlines of the Gale Crater rim in the distance, but the

WORN BY WATER? Rounded pebbles collected near Bradbury Landing (left) suggest that, as on Earth (right), they may have been formed by flowing water.

The logic driving NASA's Mars exploration programs has been "Follow the water," since the working assumption is that H_2O is essential for life.

geologists were focused on those pebbles and gravel and their most noticeable attribute: Many of them were round, or at least rounded. And as Geology 101 teaches, rocks are commonly and characteristically rounded while bumping and colliding their way down a river or stream, ending up with a look very much like the deposits near Curiosity's landing site. If it was water molding the stones at Gale Crater, then those dusty Martian pebbles might as well have been made of gold.

As recalled by Dietrich, a specialist at the University of California at Berkeley in the logic of landscapes (geomorphology) and especially the dynamics of riverbeds, "We were basically stunned. To land on top of gravel and rounded pebbles meant we had most likely landed right on top of an ancient streambed. This had never happened before, to put it mildly."

SEDIMENT INTO STONE >> On its way to Mount Sharp, as seen through a Hazcam camera (inset, opposite), Curiosity explored an outcrop given the name Shaler (above). With its inclined layers of sediment turned to stone, this formation also suggests the presence of once flowing water. On Earth, features resembling Shaler are generally formed by turbulent rivers.

For decades, the logic driving NASA's Mars exploration programs has been "Follow the water," since the working assumption is that H_2O is essential for life. That goal of finding evidence of a watery past had been greatly facilitated by the extremely high-resolution cameras sent to circle Mars since the mid-1990s, especially the Mars Orbiter Camera (MOC) and, more recently, HiRISE. Both have sent back countless images identified as ancient channels, deltas, and fans most likely formed by water. A landed mission dug up some water ice, too, in the northern polar region.

These findings and more have greatly strengthened the theory that Mars was much warmer and wetter billions of years ago, and as a result more likely to have once harbored life. But some argued the sculpted landforms revealed via satellite imaging could have been formed by geological dynamics that did not involve a warm and wet Mars, and so the issue was still open. On Curiosity's very first night on Mars, however, it became apparent that it probably wouldn't remain open for long.

While it was good luck to land on a site where water had seemingly once flowed, it was hardly pure chance. That's because the Curiosity team knew far more about its prospective landing site than on any previous Mars mission, having pored over the ever more detailed and precise images sent back by orbiting Mars satellites.

And one thing they knew was that their landing ellipse was close to the spidery remnants of an ancient alluvial fan. These fans are distinctive and common features on Earth, known to

Curiosity came across several rock outcrops
that added substantially to the now quickly
evolving water story.

be built by streams that pour water and sediment out onto slopes and down into lowlands. Exploring near portions of that fan had always been a possibility; in fact, the presence of the fan was one reason that Gale was selected as the landing site. After finding the rounded pebbles, a further look in the direction of the fan became a necessity.

Over the next few weeks, the rover was driven toward the more clearly defined regions of the fan, though never reaching it. But with continuing good fortune, Curiosity came across several rock outcrops that added substantially to the now quickly evolving water story.

TELLTALE OUTCROPS

From the high-definition color photos by then available, the pebble brigade was able to identify formations where thousands of those small rocks appeared to have been cemented together with finer-grained sands and sediment. The outcrops—the first found on Sol 27 and named Link, the second identified on Sol 39 and dubbed Hottah—were fully exposed and precisely the kinds of formation geologists need in order to read the rock history of the area. John Grotzinger, the Curiosity project scientist who is responsible for the mission's entire 400-plus science team and their work, likened Hottah in particular to tilted Los Angeles pavement.

A streambed in the Atacama Desert of Chile shows the same kind of size sorting of rocks, gravels, sand, and finer-grained sediment as identified at the fossil alluvial fan in Mars's Gale Crater.

Researchers immediately saw that the outcrops were made up of layers of rock alternately rich in pebbles and sand. Rock layers are a text to geologists, filled with clues about when the rock was formed, the environmental conditions at the time, and the relation of the layers to other outcrops in the area.

The scientific question was whether the rocks were "conglomerates"—sedimentary and formed in the presence of significant amounts of waste—or of volcanic origin, or put together by landslides or wind. Williams, a geologist with the Planetary Science Institute in Tucson, Arizona, helped lead the process at JPL and recalled the high-pressure, high-excitement, low-sleep days and nights when she and others pored over the images.

"There were lots of different views to begin with, and some wariness about identifying them as conglomerates so early in the mission," she explained. "That's a big step, and people wanted to be careful. But the evidence coming in was strong. There was a great fresh exposure on the Link outcrop, and we could see rounded pebbles in and around the rock." The shape of the pebbles, together with where they lay, made the classification of the rock as a conglomerate pretty straightforward. "Then at Hottah not even two weeks later, we found the same kind of exposure

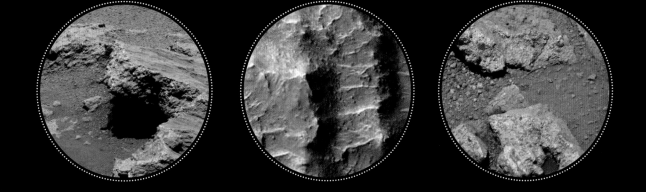

SIGNS OF A WATERY PAST >> Evidence for flowing water came to the Curiosity team early and in many forms: (above, left to right) the conglomerate rocks at Hottah, remnants of river channels, and outcrops altered in a way characteristic of past rivers or streams at Link. Because of the promise of such findings, Curiosity took a detour on its drive to Mount Sharp and traveled instead to a low-lying area named Glenelg (below). Scientists were drawn there by both the depression where water might have pooled and the triple junction—an unusual place where three different rock units come together.

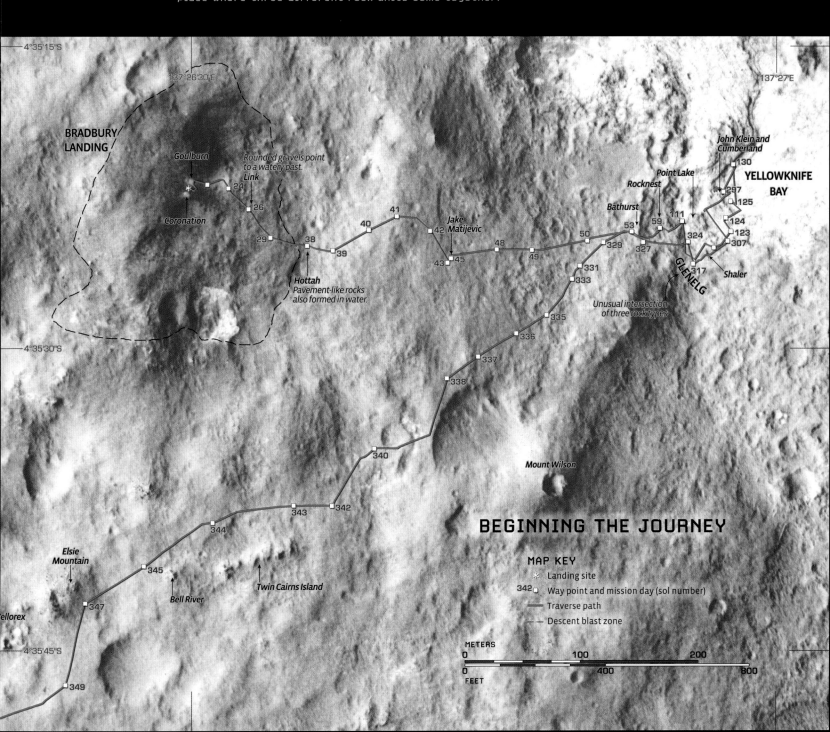

4°35'15"S
137°26'30"E
137°27'E

BRADBURY LANDING

Goulburn
Rounded gravels point to a watery past.
Link
24
26
29
38
39
40
41
42
43 45
48
49
50
Jake Matijevic
Coronation

Hottah
Pavement-like rocks also formed in water.

Rocknest
Point Lake
Bathurst
53
59
111
327
329
331
333
335
336
337
338
340
342
343
344
345
347
349

John Klein and Cumberland
130
297
125
124
123
307
324
317
Shaler
GLENELG

YELLOWKNIFE BAY

Unusual intersection of three rock types

4°35'30"S

Mount Wilson

BEGINNING THE JOURNEY

Elsie Mountain

Bell River

Twin Cairns Island

ellorex

4°35'45"S

MAP KEY
✳ Landing site
342 ▫ Way point and mission day (sol number)
—— Traverse path
– – Descent blast zone

METERS
0   100   200

FEET
0   400   800

and features," Williams said. "We weren't even in the area generally identified as the fan, but we were seeing these very clear features indicating sediment that was moved by water."

Williams, an expert in those distinctive "inverted channels" already identified in Gale through the mapping, focused on the Link images and began counting pebbles in the images of the rock. Then she dug up images of conglomerates from the Atacama Desert in Chile, a super-dry area often used as a Mars analogue. She measured and compared the size of the pebbles—ranging from M&M to golf ball—characterized their roundness, and ranked them according to the size of the rock "grains" they were made up of.

TRACING ANCIENT WATERS

Dietrich worked on where the water likely came from—a deep canyon in the Gale Crater rim called Peace Vallis—and made estimates of how much water flowed and how quickly. Even as far downstream as Link and Hottah, he and Williams concluded, the water was ankle to hip high and was once moving at one meter a second, equivalent to a normal walking pace. The size of the rounded rocks made it easy to determine they had been brought downhill by water; the Martian wind simply could not push large rocks long and far enough to cause the rounding.

All this work was done while the scientists at JPL were on Mars time—meaning, in effect, that they were permanently jet-lagged—and was often in the middle of the night. Soon after the images for Hottah had come down, the whole science team came together to discuss what had been found and to see if they could reach a consensus on what it was. A show of hands was ultimately taken to gauge whether there was a scientific consensus, and the team members overwhelmingly agreed that the rocks were conglomerates and had been formed in a river or stream.

That vote ratified the first ground-truthing ever of the presence of flowing water on the surface of Mars. That was momentous on its own, but the implications were greater still, as Dietrich explained.

"For decades, our community has looked at satellite pictures of Mars and said, 'This is a river channel,' or, 'That is a delta,' and we were pretty sure we were right," he noted. "But now we have data, proof, that Mars did have flowing water in Gale Crater, and that makes all the hundreds of other identifications of water features we've made considerably stronger. Really, it was the start of a new era of understanding water on Mars."

Williams and Dietrich presented the findings at a press conference where Grotzinger also spoke, describing the team decision about the outcrops' watery pasts as a pretty easy one. But while a new era in understanding Mars was starting, some far more difficult and contentious water issues were already being debated by the science team, and they were not considered ready for public discussion. That's why Dietrich and Williams were told not to mention at the press conference what became known as "the L-word"—*L* as in *lake*.

As would become clearer as the months went on, the initial finding that water had once flowed on the surface raised many

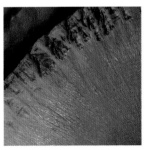

THE APPEARANCE of what looked to be a newly formed gully at the crater Terra Sirenum suggested flowing water today, though some argue otherwise.

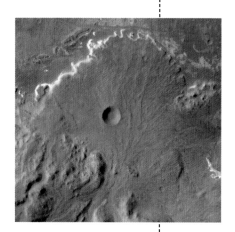

The fossil remains of a large alluvial fan appear in this image of Holden Crater, in Mars's southern highlands, taken by the visible-light camera of the Thermal Emission Imaging System. Numerous gullies worked their way down the crater walls and fed into the fan.

inevitable and highly controversial questions. First among them was whether the water flowing down from the crater rim ever pooled and formed a pond, a shallow temporary lake called a playa, or a deeper and more long-lasting lake. It would make sense if it did, since that is a common occurrence on Earth.

Researchers tended to have strong views pro and con on the subject, but the data remained insufficient to make a firm decision in the eyes of Grotzinger and others.

Much was at stake: If the long-ago existence of a Martian lake was also ground-truthed, as opposed to remaining theoretical and based primarily on orbital images, then that would rewrite important Mars geological history. It would also increase the odds a bit in favor of ancient primitive life on Mars, since lakes are considered ideal birthing grounds. What's more, the existence of an ancient lake would make it more likely that any possible ancient life or other organic material might actually be preserved in the former silt that had now turned to rock.

With the stakes so high, caution was the order of the day, and so the L-word was for private discussions only at that point. Nonetheless, it remained very much on people's minds—especially Dietrich's.

A CHANGE OF PLANS

As part of the mapping project that he, too, had joined, Dietrich and his colleagues had studied orbital images of the Peace Vallis fan and had added to the effort their own special knowledge of topography—the ups and downs of the landscape. Although they were limited to information from afar, they did have some 3-D capability and were able to put together a topo map of the fan and the area just past where it seemed to end.

On landing night, they didn't have much time to savor the presence of those rounded pebbles because they were already on to something potentially bigger: According to their topo maps, Curiosity had landed less than a mile from what was not only the lowest point of the fan, but among the lowest points in the crater. On Earth, of course, water descends to and collects in the lowest areas. The same is true on Mars.

Armed with that information, Dietrich made presentations to the entire science team assembled at JPL on both landing night (Sol 0) and then again on Sol 3, describing their most enviable situation. If they would make an admittedly major detour away from Mount Sharp and toward the depression they had found, they just might hit pay dirt in the form of the crater's terminal sink—the kind of ancient pool or lake bed that Mars scientists have hunted for decades.

There were other reasons to want to explore the unusual geology around the low point, features identified by HiRISE well before Curiosity landed on Mars.

Grotzinger was drawn to the coming together of three different units of rock identified at the low point, an unusual occurrence that is a gold mine for geologists. These contact points are where geologists can learn which rock units were laid down first, which came next, and all that implies. What's more, one of the rock units was light toned and showed what scientists

TRACING ANCIENT WATERS >> Rebecca Williams (right), a geologist and member of the Curiosity team, helped identify the fossil fan named Peace Vallis within Gale Crater. She and her colleague Bill Dietrich led the effort to piece together specifications of the long-lost flowing water.

Williams also studies inverted channels, vestigial ridges of sediment cemented into rock and left exposed when the rivers dried up and looser material eroded away. The inverted channels in Mars's Kasimov Crater (below) are especially well defined.

 ⌄ AUGMENTED REALITY
View this image through
NASA's Spacecraft 3D app.
Learn more on page 6.

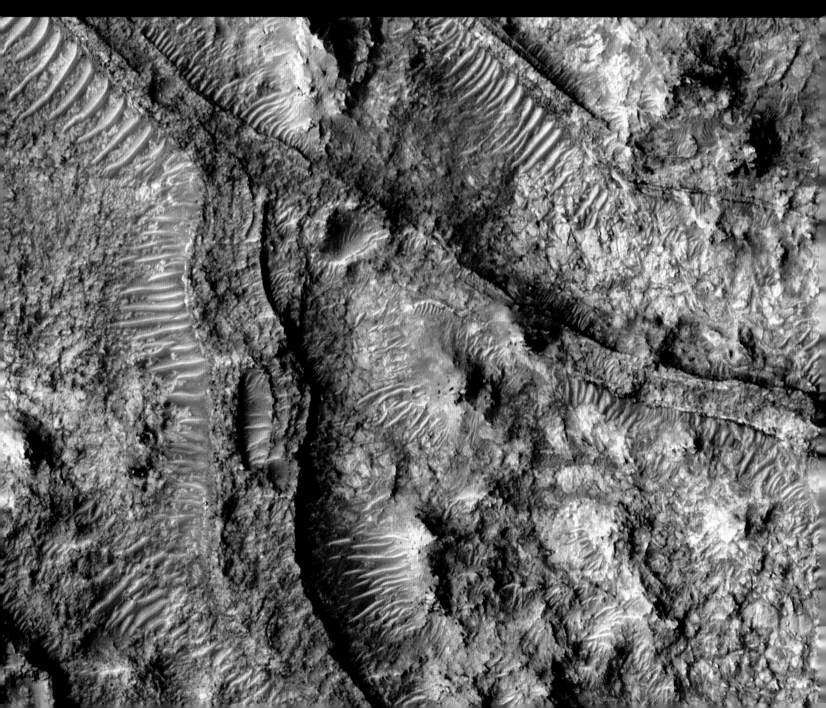

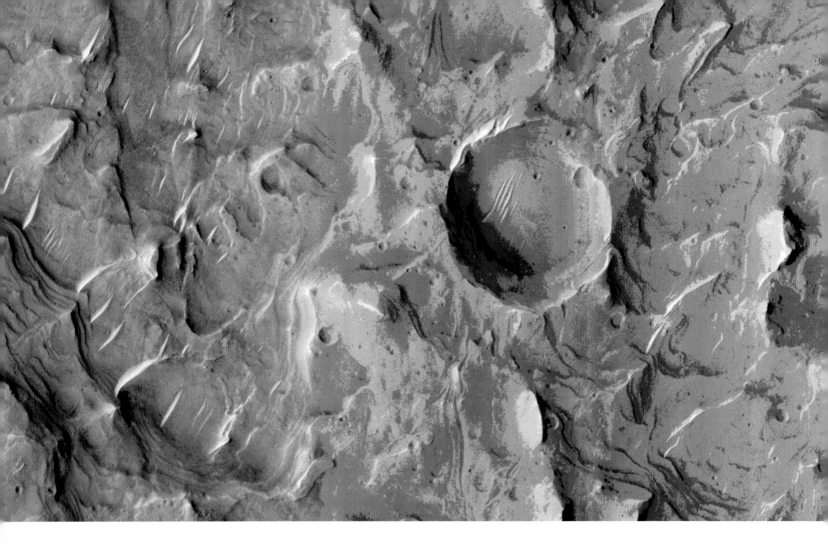

describe as "high thermal inertia," the physical characteristic of keeping ground heat and ground cold longer than nearby bedrock. Formations with low thermal inertia are often associated with the fine clays washed down by streams and rivers.

So the possibility of finding an ancient lake bed and then drilling into it was the big draw for sure—and was the primary goal of a group of advocate scientists (led by Dietrich and Williams) who called themselves "the fanatics." But the project scientist and others had their own reasons for redirecting Curiosity; this was formally a "mission of discovery," and there was potentially a lot to discover at the low point.

Without Grotzinger's support, the rover probably would have headed directly to Sharp. With it, the team voted on Sol 11 in another show of hands to change course and head in a direction opposite to long-established plans.

THE DETOUR

Initially, Curiosity leaders expected the trip to the low point—given the palindromic name of Glenelg in part because the trip was expected to be a quick there-and-back—to take several weeks to two months. A minor detour in many minds, but the plan nonetheless carried real risks.

One of the primary goals of the mission was to reach Mount Sharp, the geologically rich Promised Land of Gale Crater. While Glenelg was not far from the landing site, going there was a deviation that would be argued and criticized for years if anything went wrong or the science turned out to be inconsequential.

An alluvial fan formed inside a large unnamed crater in the southern hemisphere of Mars: Such fans form when water and small bits of sediment sweep down crater walls and spread out. This picture was taken by HiRISE and enhanced with false color.

>> Whitish deposits, fans, and channels show clearly in an image of Mars's Ladon basin: ancient lake or river sediments, perhaps? Some of the light-toned material may be clay, good for preserving evidence of organic material.

That left only one other alternative:
the quite radical notion that Mars once
had widespread precipitation.

So the detour could have turned out to be a huge mistake. But as we'll see, the team's presentations and predictions in those earliest sols represented very good science indeed.

The L-word became a focus for endless scientific debate and chatter among the Curiosity scientists; more than one of the songs, poems, and ditties delivered at the daily JPL update and planning meetings of the first 90 days on Mars involved hopes and dreams of finding a lake. But the excitement it kicked up about water on Mars quickly led to other basic and entirely unresolved questions about the planet's aqueous past—the Martian big picture.

For instance, the next obvious question to ask after how and where the Peace Vallis fan ended is, Where did the water flowing down into Gale originate? Most immediately, it clearly came from a highlands that also had some ancient riverbeds. But where did *that* water come from?

The possibilities were pretty limited—but again potentially of enormous importance in terms of writing the history of Mars. Researchers have traditionally proposed that groundwater either bubbled up from beneath the surface of Mars, came from melting glaciers, or was triggered by heat-producing impact events.

"I'm a topoholic. Things make topography, and every elevation difference tells a different story."

William Dietrich, University of California, Berkeley

A leading geomorphologist, Bill Dietrich brought years of studying the evolution of Earth landscapes to Gale Crater. He studies environments from a topographic point of view, as in this early analysis (below) of Peace Vallis.

Dietrich's research has focused on important but little-understood subjects: how sediment flows down a river or mountainside; what determines the spread of water in a floodplain; what dry channels can tell us about a landscape. His floodplain work has taken him many times to the Fly River in Papua New Guinea, where sediment waste from a large mining operation upstream allows for precise tracking of water flow. His expertise has environmental implications on Earth and has been important to understanding Gale Crater and Mars as well. Dietrich and Curiosity colleagues have conducted a detailed analysis of the Peace Vallis alluvial fan, which has led him to a surprising discovery: There has been so little work done to understand these fans on Earth that soon we'll know more about them on Mars than here.

The bubbling-up theory has had many proponents over the years, since it is known that huge deposits of water ice are buried beneath the surface. A volcano, a large meteorite, or a warmer era when a Martian pole directly faced the sun during a period of extreme tilt could all theoretically cause polar and underground ice to melt and flow on the surface. But the near-equatorial location of Gale makes that interpretation less plausible, since the polar water would have had to travel so far. The impact and subterranean-melt theories have proponents, but data are scarce.

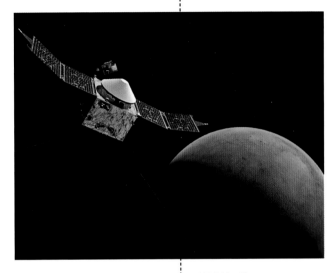

That left only one other alternative, and it was being embraced by an increasing number of Mars researchers: the quite radical notion that Mars once had widespread precipitation—a water cycle with rain or snow, a pooling of the falling water, evaporation that formed clouds, and then more precipitation again. The image seems almost surreal: a rainy day on Mars, maybe even with thunder and lightning in the distance.

Or, because Mars is farther from the sun and inherently cold, the water cycle could have involved just snowpack and glaciers. They would melt when meteorites struck, volcanoes erupted, or other unusual dramatic events took place.

ANCIENT HISTORY

Another question that inevitably arose from the finding of once flowing water asks when in Mars history the river would have existed. Dietrich believed that it may have lasted for thousands to millions of years, but he had no way yet to date it.

Nonetheless, a consensus had grown that the water almost certainly flowed after the crater was dug by a large meteor impact: In other words, the impact had not exposed an ancient, covered-up river system.

The Gale impact has been dated to around 3.6 billion years ago. Traditionally, this is viewed as the boundary line between the Noachian and Hesperian eras on Mars, a time when the earlier wetter and warmer period had given way to colder, drier, volcanic, and generally more harsh conditions. Yet the river system clearly started after the crater was formed, which means that substantial water flowed into Gale Crater well after that impact 3.6 billion years ago. This scenario would certainly suggest that Mars was still relatively wet and warm at a time that most geological histories of the planet say it was cold and dry. Or, at least, that it was relatively warm and wet around Gale Crater.

What's more, Mars would have needed to have a much thicker atmosphere than it has today to keep any possible surface water from quickly drifting away. While that remains possible, there is scant evidence that the Martian atmosphere was ever thick enough to hold water. Yet, apparently, it happened.

And finally, the most confounding issue of all: How could Mars have ever had water on the surface in the first place?

Climate modelers have long argued that the atmosphere of early Mars was colder than it is now. This understanding harks back to the "faint young sun paradox" first put forward by Carl Sagan and his colleague George Mullen in the 1970s.

By all known accounts of star formation, our sun must have been considerably less luminous when the Earth and Mars were formed. That was some 4.5 billion years ago, when much smaller planetesimals began to come together in the debris disk that surrounded the sun, and both proto-Earth and proto-Mars grew to the size of planets. Based on observation of similar stars at earlier points of their development, back then our sun would have been putting out only 70 percent of the radiation (and heat) that it produces now.

Sagan and Mullen focused primarily on what the faint young sun meant for ancient Earth history—that at a time when the planet should have been a giant snowball, it instead featured oceans and rivers and lakes. Scientists have subsequently concluded that carbon-based greenhouse gases in the atmosphere might explain why Earth was warm when it should have been frigid, and now the same paradox is playing out on Mars—but on Mars captured in the rock of a similarly large greenhouse gas presence eons ago. By faint young sun measures, ancient Mars, an icebox with little greenhouse gas to hold in heat, should never have had any liquid surface water. Other explanations for the apparent warmth—large-scale volcano eruptions, other major meteorite impacts, maybe even an unusually large early sun—all pose both theoretical and practical problems based on the information we have.

So even with the growing evidence of a once wetter and warmer Mars, the faint young sun issue remains a major obstacle to understanding the water story of the planet: The climate and atmosphere modelers simply cannot, with the data and projections they have, make Mars the place geologists and planetary scientists now believe it once was.

A CONFLICT BETWEEN EVIDENCE AND THEORY

The primary warming agent in the Martian atmosphere would be carbon dioxide, and the modelers have consistently said they cannot come up with any way to produce a Mars where the greenhouse effect could bring the temperature above freezing except for limited times at limited sites. This has led to no small amount of frustration on both sides.

Bob Craddock, a geologist with the Smithsonian's Air and Space Museum, thinks it's time for the climate modelers to accept that something is off with their work.

During a science conference talk, he described Martian conditions that produced extensive river valley networks where large bodies of standing water filled craters, where rain or snow seemed to have fallen over a long time, and where enormous megafloods occurred because the ground was so saturated with liquid water. He said after the talk that he has grown weary having Mars climate modelers tell him his analyses—based generally on geology—are misguided.

"They say, 'Can't do it, can't do it—can't get the surface temperature in a carbon-dioxide-rich environment up above freezing.' My argument back is, Well, too bad. We have empirical

The Shaler outcrop
shows signs of lay-
ered cross-bedding,
an indicator for
past river activity
that molded small
underwater dunes
and moved them
downstream. This
image was taken by
Mastcam on Sol 120.

⌃ AUGMENTED REALITY
View this image
through NASA's
Spacecraft 3D app.
Learn more on page 6.

data that suggests your theoretical model is wrong. So how can you be standing there telling us that we're wrong? They have to go back and twist the knobs and tweak the measurements and return with something different because, well, we have the data and we're getting more all the time."

This kind of conflict happens all the time in science and has a well-known precedent regarding Mars and water. After the Viking landings in the mid-1970s, climate modelers used measurements of lighter and heavier isotopes of the gas argon to make predictions about how much water had once been on the planet.

The high-profile paper concluded that Mars was always quite dry, with only enough water at its peak to cover the planet to a level of 23 feet. That would have to account for all the water now frozen in the polar caps, that might be under the surface in other regions, that became locked into Martian minerals, and that flew off into space.

But subsequent discoveries about water on Mars, beneath its surface, and that has drifted off the planet undercut the 1977 paper entirely. Now, Craddock notes, the scientific consensus is that Mars once had global water coverage—all the H_2O present in all forms and at all locations—of a mile or more.

Stephen Clifford, a specialist in Martian water with the Lunar and Planetary Institute, is actively involved with Mars climate modeling and understands the frustrations of the Mars

The dynamics of alluvial fans were suddenly urgent questions.

geologists. But given the constraints put on Mars water by the faint young sun and its thin atmosphere, he is now convinced the climate modelers are wrong again. Perhaps the liquid water appears only periodically, he says, and is the result of local phenomena such as volcanoes, meteorites, and glacier melts. Climate experts look to these explanations, he says, because it remains impossible to model an early Mars that had Earth-like conditions on a global scale, and where an ongoing water cycle brought rain, snow, evaporation, and clouds across the planet.

On this global scale, Curiosity could bring more clarity, as it does with local landscapes. Clifford says, for instance, that if the rover finds that the interior of Gale once had a lake (and especially a large lake), then that strengthens the geologists' case. The same would be true if the rover team finds large and widely spread deposits of the kind of minerals that can only form in water. These features are associated with long-term exposure to water, and that would probably mean persistent rain or snow.

"The climate modelers would have to catch up fast if that's the case," he said. "And if it's not, the geologists will have to reexamine some of their assumptions and conclusions."

With all this in the background, Curiosity scientists have spread across the world to find possible analogues to Peace Vallis and other water features in Gale. Dietrich and Williams met with others in Death Valley to study shallow, sloping alluvial fans; Williams was awarded a grant to do similar work in Chile's Atacama; and Sanjeev Gupta, a colleague and co-author of Dietrich's from Imperial College, London, headed for Cyprus.

The dynamics of alluvial fans—how they form channels, how to determine their water flow, the effects of slope and debris, the possibilities for placing them in time—were suddenly urgent questions.

So was the search for information about and locations of "fan deltas." In geological terms, an alluvial fan peters out into soil, while a delta spills into a lake or other body of water. A fan delta is a kind of cross between the two, with attributes of an alluvial fan but a watery endpoint like a delta. An alluvial fan tails off, becoming ever more flat, while a fan delta has a dip before it empties into water.

Dietrich had another hunch about the terrain that Curiosity was exploring, and it involved that final dip of a fan delta. The end of the Peace Vallis fan seemed to have such a dip—but he needed to know a lot more about fan delta features on Earth before he could argue the fan delta case on Mars. And if he could make the case, then the absolute low point, nearby Glenelg, would look more and more like the L-word.

With so many big questions on the table because of the Peace Vallis discoveries and what might be found at the apparent endpoint of the ancient riverbed, the plan to quickly get into and out of the area was soon abandoned. Weeks turned into months, and the detour gradually became a destination.

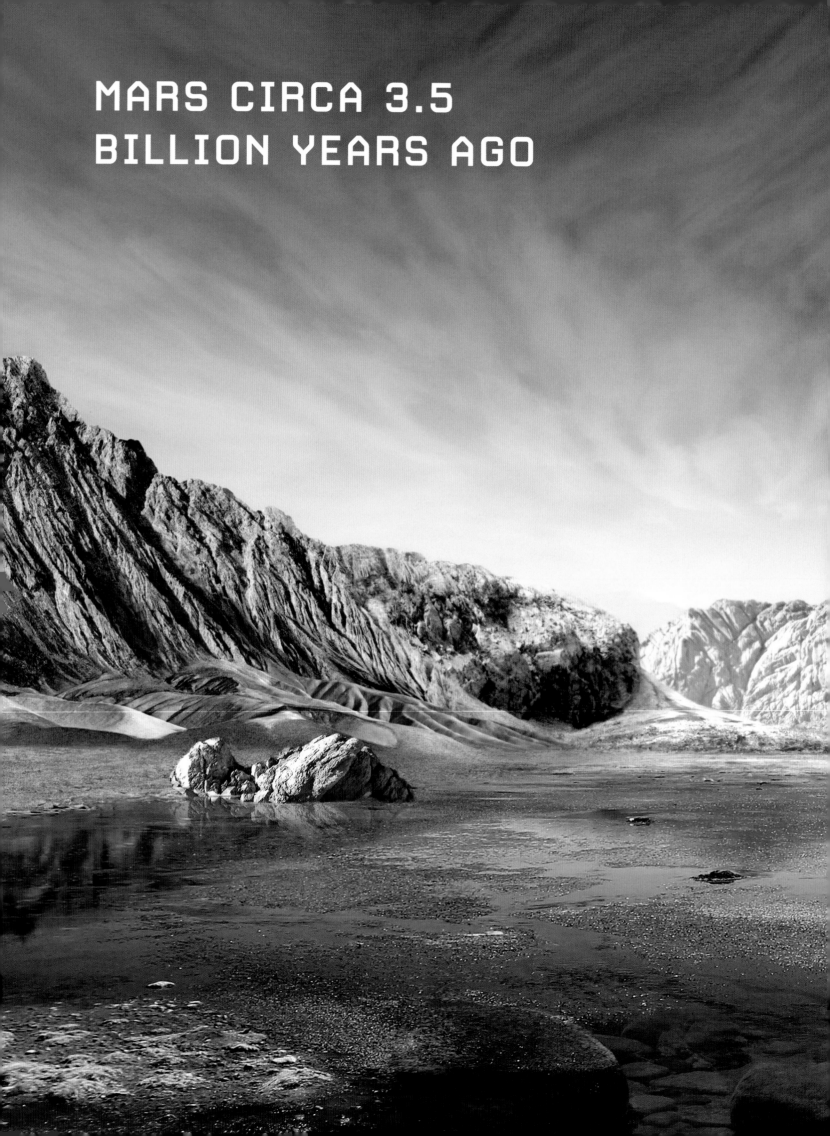

MARS CIRCA 3.5
BILLION YEARS AGO

Scientists are convinced
that water flowed for
centuries into Gale Crater.
The waterway that etched an
alluvial fan as Mars dried out
may have looked like this.

MARS LOOKS DIFFERENT

CHAPTER 5

The amazing array of imaging equipment now capturing pictures of the planet's landscapes and surfaces

EVEN BEFORE LANDING, it was clear that Curiosity would be sending back a far more dramatic and vastly more sophisticated picture of Mars than anyone had ever seen before.

The 17 cameras taking photos are absolutely essential science tools, scouts for the rover team as well as enablers of scientific discovery. And they're also the eyes that allow the rover to keep on track and not topple off a cliff or get stuck in a ghost crater.

With the high-speed descent still in progress, the heat shield was set free and the Mars Descent Imager—attached to the bottom of the Curiosity rover—started taking pictures. The result was a stop-action, in-focus, and quickly iconic image of the heat shield careering down to the surface—a flying saucer photo for real, and the first of its kind on Mars.

Soon after came images precise enough to tell Dietrich and Williams that there was riverborne gravel at the landing site. Then in the next days and weeks came images that revealed intriguingly layered outcrops of rock; others that showed Martian rocks split open by Curiosity wheels and featuring grayish blue insides; shots of millimeter-size grains of sand; and, of course, Mount Sharp beckoning in the distance.

Some were taken by a telephoto on the mast that stands atop the rover and can distinguish a basketball from a

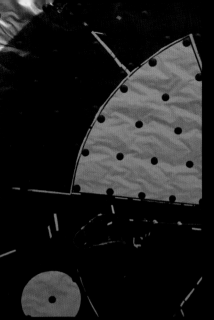

SEPARATION IMAGERY >> Tucked into the bottom of the rover, the Mars Descent Imager camera [MARDI] captured images of the heat shield as it detached from the spacecraft and fell toward Mars. The creation of space photographer and geologist Michael Malin, MARDI took the first photos of their kind. MARDI survived landing and continued to take pictures of the ground, including some intriguing views of rocks broken by the rover's wheels.

football at a distance of seven football fields. That resolution is three times greater than any landscape camera used on Mars before. Another color camera atop the mast is a broad-context, medium-angle lens designed to capture panoramic sweeps of the Mars landscape. Both the telephoto and medium-angle cameras are together called the Mastcam.

The highest resolution images came from the Mars Hand Lens Imager (MAHLI), a camera similar to the Mastcam but attached to the end of the rover's arm, that can focus on objects up close (as near as 0.8 inches, or two centimeters) as well as very far away. Geologists never go into the field without a hand lens to examine the size of rock grains and sands and to help identify textures and minerals. Now a Mars rover had an extremely powerful hand lens as well.

But it was on Sols 84 and 85 that the unique capabilities of the Curiosity cameras were put to a test never before possible. The rover was commanded to take a series of 110 pictures of itself, a full-frontal stereo portrait that would ultimately make it appear that someone—or something—was standing before it and taking the picture.

Choreography for the maneuver had been worked through on the JPL test bed just before landing. The photo session on Mars involved a two-day dance by the seven-foot robotic arm and the MAHLI camera, a stand-in of sorts for a geologist's magnifying tool, that sits at the end of it.

To make the maneuver even more complicated, the camera lens had to be kept in one place as much as possible to minimize parallax—the distortion of images caused by a change in camera position—while the arm went through its contortions.

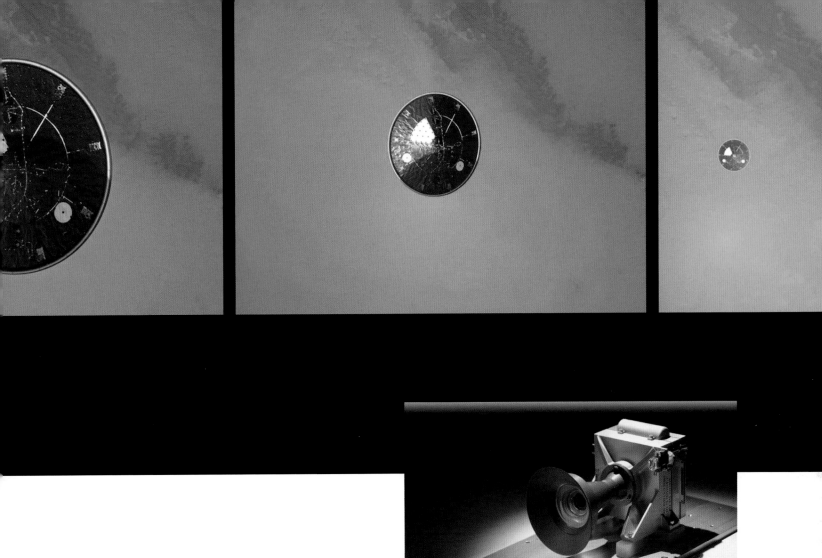

>> The MARDI camera, not much
wider than a pocketknife, takes
four frames per second and therefore created a near-video record
of the final descent at 1,600 by
1,200 pixels per frame.

Once the rover transmitted the images back to Earth, the MAHLI team (and hobbyists around the world, as the data were immediately released on the Internet) set out to compile a single mosaic image out of them.

There were challenges galore, but none quite so great as getting the incompletely imaged bits and pieces of the rover's long arm out of the picture. A close look at the final image shows the shoulder of the robotic arm on the front of the rover. But the rest of the five-jointed appendage had been removed and replaced by pixels from other frames that showed the ground or the rover behind it.

The picture demonstrated the new capabilities of the rover and gave the vehicle a face, a personality, and a reality that only a photo can deliver. The self-portrait became an instant hit, and both it and a JPL animation video made of the picture-taking process are now essential to virtually all storytelling about Curiosity on Mars.

FIRST SELF-PORTRAIT >> Using the MAHLI camera at the end of the seven-foot robotic arm (left), the Curiosity rover has taken a series of unique self-portraits. This version (above, left) was the first and was taken at Rocknest, the vehicle's first major science stop. The arm choreography needed to take the photo is complex and was practiced numerous times on the Curiosity twin at JPL (above and at left) before landing. The commands are written and sent to the vehicle by rover planners, better known as rover drivers. The finished product is a mosaic of dozens of images taken on Sols 84 and 85, then reworked slightly back on Earth to take the robotic arm out of the photo. Look carefully and you can see the turret at the end of the arm, as well as the Martian landscape, in the round reflective ChemCam instrument atop the rover's mast.

NOT ALL RED

The image jumps out not only because of the drama of seeing the rover full face and in place on Mars—a theatricality due in part to the involvement of filmmaker James Cameron. (Yes, a 3-D version will be coming later.) The image is also crisp and devoid of the reddish dust that is omnipresent in earlier images from the surface of Mars and its thin atmosphere. Removing the color-muting dust in the atmosphere to the extent it's being done—through a process called white balancing—is another Curiosity innovation. The cameras have broader bands designed to specifically remove the dustiness, resulting in color images that look the way our eyes expect. Previous Mars surface cameras took photos through colored filters that were then combined back on Earth.

In the end, the butterscotch color created by the Martian atmosphere is gone—just as white balancing takes the greenish hue out of photos taken under fluorescent light and the yellowish cast that comes from incandescent tungsten light. Instead of a muddied view with few sharp features, you see an image that matches what the eye would see if the scene were on Earth.

This intensity is additionally difficult to achieve because Mars is considerably farther from the sun than Earth—on average 141 million miles versus 93 million miles, and that means less light arrives to the Martian surface. The amount of light is less than 50 percent of what hits Earth—making the brightness of Martian noon seem like late afternoon on our planet. It never gets particularly bright on Mars.

A MASTERMIND OF MARTIAN IMAGING

Several different teams of researchers and engineers can claim credit for the revolution that has taken place in photographing Mars, where orbiting cameras can produce images using a billion pixels and that have resolutions—from 50 miles up—of one pixel per square foot.

But there is one acknowledged master in the field of Mars photography, and his name is Michael Malin. He is a tireless and laser-focused visionary, by all accounts brilliant, if sometimes reclusive and a tad cranky. He works out of a low-slung office building in an undistinguished industrial park in suburban San Diego. The building doesn't even have the name of his company anywhere to be seen; you have to know where you're going in order to get there.

This is the home of Malin Space Science Systems, founded by Malin with the proceeds of his MacArthur "genius" grant in 1987. That so much essential and revelatory Mars camera work depends of the efforts of this man and his staff of 30, as opposed to a much larger corporation, is remarkable. Malin has been active in photographing Mars for nearly every NASA mission there since Viking in 1976, and has taken images for non-Mars missions too.

The technical wizardry and scientific accomplishments of Malin and his colleague Ken Edgett are well known to the Curiosity team, but far less so to the public. Even fewer outside the Mars community know that Malin not only develops ever

 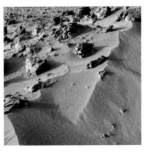

COLOR CORRECTION New imaging processing techniques change a dusty image [left] by white balancing to make it look as if it was taken on Earth.

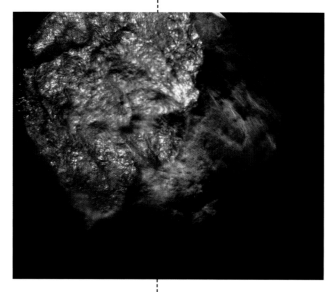

The ChemCam camera is generally used in conjunction with the instrument's laser and spectrometer to analyze elements in Mars rocks. But its high-resolution fish-eye lens can also take unique black-and-white pictures of objects as small as this tiny pebble or as large as Mount Sharp. Here, color has been added from Mastcam images of the same target.

more capable and versatile cameras, but that he and Edgett (both trained as geologists) also wrote numerous pioneering articles in *Science* that described sedimentary layers on Mars and, later, possible gullies with water that may still be flowing. What's more, they provided much of the early and very promising information about the rock layering at Gale Crater that ultimately led NASA to select it as the landing site.

But Malin and Edgett's driving ambition with their four cameras on Curiosity that they provided and manage is to deliver images of Mars as they would be seen and processed on Earth. This has led many to assume the images are "colorized," but Malin said the process is far more nuanced.

"It's no simple thing to say what the right color is," he explained. "Color is a very, very complicated thing.

"When you ask the question 'What would Mars look like if you were standing there?,' the answer involves how the human eye and brain have evolved to interpret and understand color. And the answer involves as well whether you're talking about what your eye sees as compared to what colors a camera would capture."

What he and his team are doing is both providing the raw images of a butterscotch Mars, and then the more processed images that correct for all the factors that keep them from looking Earth-like. The result is a library of Mars images that are unique, viewer friendly, and that contain, in Malin's words, just a touch of art. They allow us to see Mars "in the light our eyes evolved in," as Malin put it.

NEW REALMS OF PHOTOGRAPHY

The implications for Mars science are enormous. Curiosity cameras can identify rocks that have unusual colors and, consequently, might have surprising component parts. And they can bring the entire landscape to life far better, pointing geologists to the spots where interesting rock layering has occurred, where different rock units meet, where the bedrock is fractured or worn by water.

What's more, filter wheels enable both Mastcam cameras to distinguish near-infrared wavelengths, and the ratios of brightness they pick up when "photographing" a rock can be used to spot the presence of minerals formed in water. This did indeed happen later when the rover was first scouting the depression near Glenelg.

And they can identify unusual patterns and colors worthy of future investigation. On the way to the basin, the Mastcam took a picture of an outcrop that seemed odd, somewhat out of place. The rocks were given the name of Shaler, and the team determined to give it more attention when returning from the low point.

Some 180 sols later, that's exactly what happened, and Curiosity did a far more extensive examination of the outcrop. The result was the discovery that Shaler was textured, layered,

"Mars has never been real to people because they haven't been able to see it like they see places on Earth. My goal is to make it real."

Michael Malin, Malin Space Science Systems

Principal investigator for three Curiosity cameras, Malin has provided many images that redefine Mars, such as this photo (below) of Mount Sharp. His team specializes in areas ranging from color properties of objects to cold climate landscapes.

It is said that Malin has seen more of Mars than anyone else on Earth. He has been involved with virtually all Mars missions since Viking, and has pored over images and other readouts of Mars—many taken by his instruments—for decades. This has resulted in numerous breakthrough science papers, often written with his colleague Ken Edgett. His cameras have helped deliver highly detailed understanding of the surface of the planet. Malin and Edgett, for instance, were the first to describe in great detail the sedimentary nature of some regions in Mars. Prior to their work, the planet was considered to be largely volcanic and absent the transformative sediments that are moved around by wind and water. Years passed before their view of a sedimentary Mars was fully accepted, but today the Curiosity mission is an exploration of primarily sedimentary rocks.

16.2 KILOMETERS
Ridge top

9.3 KILOMETERS
Layering of buttes
tells sedimentary
story

6.6 KILOMETERS
Crater remains,
with surrounding
boulders

5.5 KILOMETERS
Buttes in
black sand

3.7 KILOMETERS
A depression,
or swale

125 METERS
Gravels near
landing site

<< This view of Gale Crater and Mount Sharp, taken as a test series to calibrate the 100-millimeter Mastcam, shows the gravelly foreground, a depression (or swale) beyond that, and then the black sands below Mount Sharp. The top ridge is 16 kilometers (10 miles) away.

>> Ken Edgett is the principal investigator for MAHLI and, with Mike Malin, uses photogeology to produce pioneering Mars science. Here he speaks to colleagues at JPL, the image behind him fuzzy due to dust on the camera lens.

and made up of components found all around the bowl below. In other words, Shaler strongly indicated that all the rock layers nearby were related, and that's a key finding for geologists. It tells them when in time layers of Gale Crater rock were laid down, even if the layers aren't close up against each other.

Curiosity's camera capabilities are great, but they were initially planned to be greater still. Malin, along with colleague Michael Ravine and director James Cameron, proposed and were approved to build zoom capabilities into the two Mastcam cameras, an advance that would have greatly expanded Mars 3-D photography. The project was in its later stages of design when it was descoped from the mission, dropped because NASA managers said the technology was not sufficiently developed—a conclusion that Malin disputes.

The Mars Descent Imager, which took photos from the capsule's back shell on landing, also almost didn't make it on board. The issue there was cost, and what was described as the priority level of the project. Faced with the possibility of that camera also being descoped, Malin initially paid to finish the work out of his corporate funds, then paid his staff to calibrate the camera before delivering it to JPL. The opportunity was just too great for a man who thrives on challenges.

"How could MSL land on Mars without this camera to actually, for the first time, shoot the descent?" Malin asked. "This was not a possibility we were going to let pass by."

The fuzzy contours first photographed by the Mariner flybys in the 1960s showed the outlines of an alternately cratered and uncratered surface, but the higher-resolution cameras on the Mariner 9 orbiter in 1972 hinted at a Mars with a potentially intriguing geological past. Valley networks for rivers and streams were identified along with some large channels, and they raised hopes that a lander might just find signs of life.

Surely there would be no vast canal systems dug by intelligent Martians, as some 19th-century astronomers thought they saw on the planet. But something more modest just might be found that nonetheless met the definition of being alive.

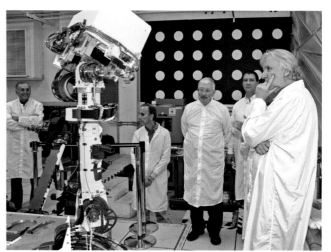

A composite of two photos taken during the
1997 Pathfinder/Sojourner mission creates a
unique view of sunset on Mars. The sky is a
true-color image, while the terrain in the foreground was adjusted to match the time of day.

>> Men in white coats—including filmmaker James
Cameron (a member of the Mastcam science
team), JPL director Charles Elachi, project scientist John Grotzinger, and Michael Ravine of
Malin Space Science Systems—watch an engineering model of Curiosity's mast cameras in 2010.

When the Viking landers touched down in 1976, however, they sent back images of a
parched and desolate Mars. The planet looked to be dominated by endless plains pocked with
craters, and the entire landscape appeared reddish brown: the air, rocks, and Martian soil all
colored by that rusty dust. Testing for life-forms and organic materials was inconclusive, but
generally interpreted as negative. Many Mars scientists now see the Viking mission as grand,
but also scientific overreach.

Much of the public was left primarily with those bleak images, and they came to define
the planet. The notion that Mars had hidden secrets—that it had a past quite different from its
present—was a tough sell for years to come.

Thanks to exponential advances in the resolving power of cameras built and deployed by
Mars scientists and engineers in recent years, it is now clear that both earlier views are incorrect. Through the lenses of orbiting satellites and now Curiosity, the landscape has changed
again—and for good.

Mariner 4 was the first spacecraft to send back close-up images of Mars, but the speed with which they were being transformed from digital data was not fast enough for some. In 1965, using the numbers printed on strips of paper as their guide, NASA scientists color-coded the information (left) and used it to make a paint-by-the-numbers image of Mars (above) that was presented to JPL director William H. Pickering.

There certainly are desert highlands and lowlands, as well as untold numbers of craters and some ancient volcanoes. But the super-high-resolution cameras on orbiting satellites have now shown with much greater precision that Mars also was sculpted by long-ago flowing rivers and possibly sizable lakes.

It has deep packages of sedimentary rock and substantial fossil river deltas. It also has hundreds of the alluvial fans identified from above throughout Mars and—in the first major scientific finding made by Curiosity—now confirmed on the ground at Gale Crater. Scientists know all this because of cameras, which have become an essential tool for conducting Mars research.

Malin's Mars Orbiter Camera (MOC) on the Mars Global Surveyor satellite (1997–2006) advanced the science of Mars photogeology by giving form and meaning to hitherto indefinable features. It was the MOC that first identified Gale as an especially compelling site by identifying the extensive layering at the base of Mount Sharp.

As Edgett tells it, the discovery was fortuitous because in its early years, the MOC could not be pointed at features of interest; scientists had to wait for features to come into view, directly beneath the orbiter.

Pointing would come later. To do that with the MOC, the entire Mars Global Surveyor satellite had to be rotated to allow the camera to target, and that was initially considered a risky and not fully certified maneuver.

During the prime mission of the satellite—the two years after the satellite was in proper orbit that NASA defined as a successful venture if completed—the camera was allowed to target but twice. The first time, Edgett recalls, was to take images of the formation known then as "the Face on Mars." The second was to search the area where the Mars Polar Lander was believed to have crash-landed in 1999. (The spacecraft has never been found.)

So Edgett and Malin were delighted when the prime mission was completed and their campaign based on pointing the camera began. They had already collected the data that led to the revolutionary finding that sedimentary rock covered many areas of Mars—a discovery that overturned the consensus at the time that the surface was primarily covered by unaltered rock spit out by volcanoes. Because sedimentary rock needs water (as well as wind and glaciers) to break up the original volcanic bedrock, that means Mars has long been a geologically active, rather than static, planet.

But with the pointing they could turn to specific features of interest, and in particularly river deltas. A paper on the Eberswalde delta, the sculpted remains of a large delta that looks very Earth-like, followed.

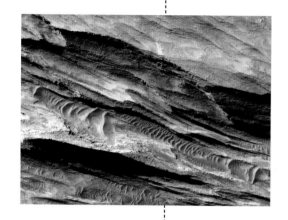

The HiRISE camera highlights concentric rings of sedimentary layers in Mars's Ceti Mensa basin. Dark sand ripples and textures in the bedrock speak of scouring and erosion caused by winds. The combination of orbital imaging and Curiosity photography allows scientists to identify and analyze potentially important phenomena.

THE HIRISE EFFECT

The next giant step was taken by the 143-pound High Resolution Imaging Science Experiment (HiRISE), flying on the Mars Reconnaissance Orbiter (MRO).

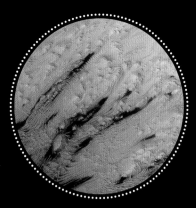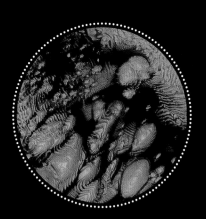

ENHANCING OUR VIEW OF MARS >> A series of images of Mars's West Arabia Terra Crater shows how the initial image (above, left), taken by the Mars Orbiter Camera on the Mars Global Surveyor satellite, can be enhanced by the addition of color to bring out the contours and features of the planet. MRO (the Mars Reconnaissance Orbiter, below) is home to HiRISE (the High Resolution Imaging Science Experiment) and CRISM (the Compact Reconnaissance Imaging Spectrometer for Mars), a spectrometer that identifies important mineral deposits around the planet. MRO orbits between 155 and 196 miles (250 and 316 kilometers) above the surface and has produced a vast catalog of Mars images, some of them abstract, surprising, and beautiful.

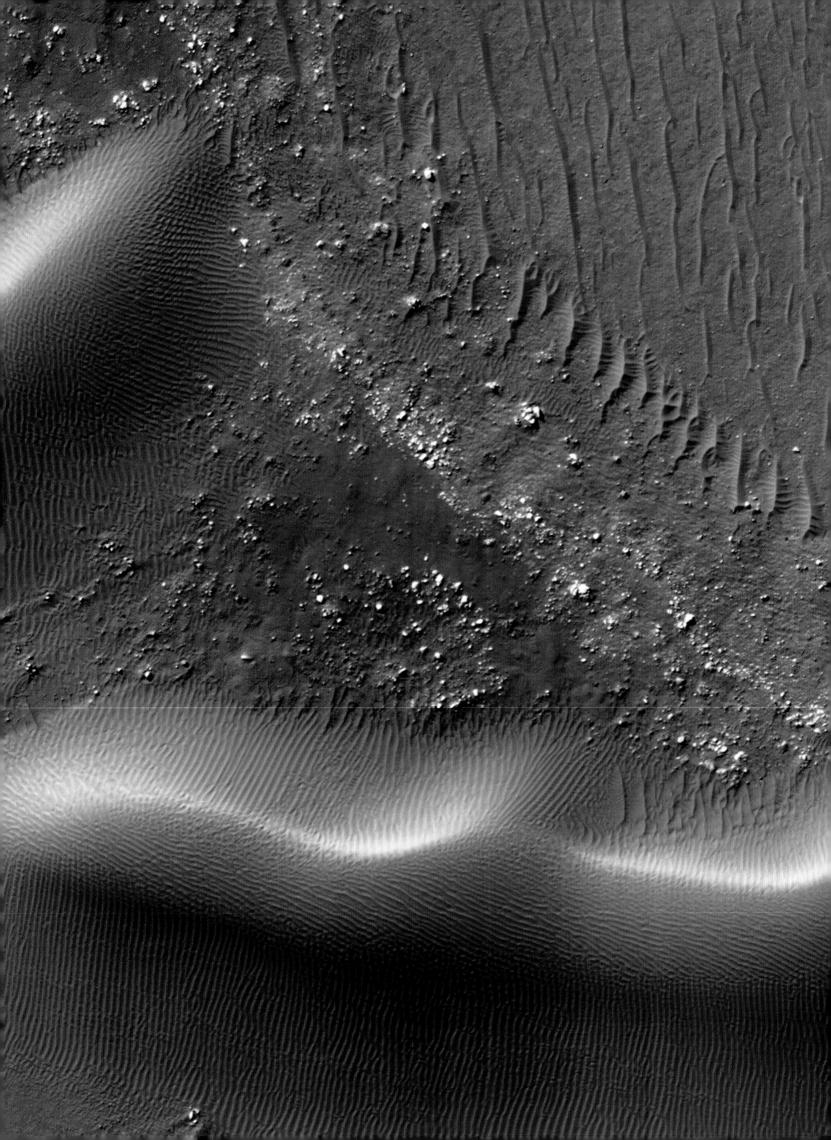

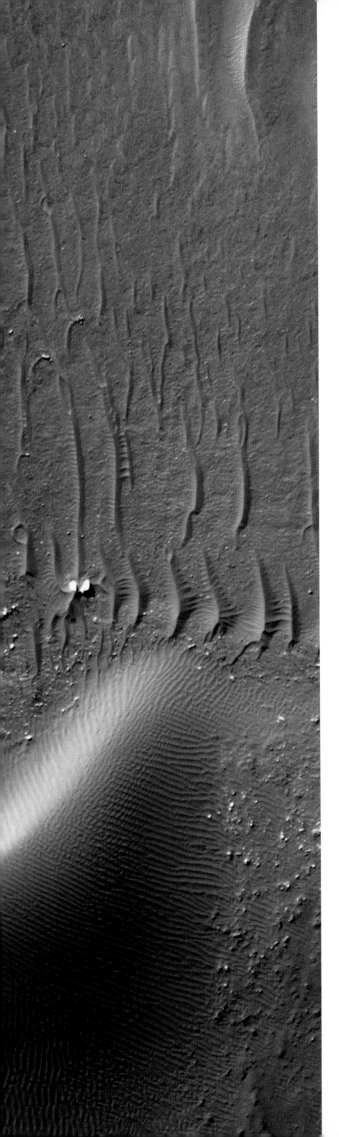

Turned on in 2007, it could identify and image an object the size of a cow on Mars from its orbit 200 miles above the surface. It could achieve this resolution—about one foot per pixel, versus 4.5 feet per pixel on the MOC—thanks to a telescope with an aperture of almost 20 inches connected to a large, CCD digital camera.

With that level of detail, NASA has been able to scout and map potential landing sites far more extensively. HiRISE is now surveying locations for the planned 2016 Insight mission (a lander to study the deep interior of the planet) and the scheduled 2020 reprise of a second Curiosity-like rover. Once Gale had been selected as the current rover's new home, the crater became what is believed to be the most photographed place on Mars.

The impact on Mars science has been equally profound. The high resolution confirmed Malin and Edgett's conclusion that layered sedimentary rock is surprisingly common across Mars, and expanded the inventory of alluvial fans, deltas, gullies, and other water features. HiRISE takes pictures that come back largely black-and-white, with indistinct features. But it can digitally paint the features below with colors that can pull off the dusty cover of the landscape and show details otherwise hidden—an effect parallel to Malin's white balancing.

SEASONAL STREAKS

Among the top HiRISE discoveries is that what had been hypothesized to be briny water regularly appears on some steep crater slopes. During the coldest months the walls are featureless, but when the Martian summer arrives they have many and long streaks that certainly look like some kind of flowing material.

The seasonal streaks (officially known as recurring slope lineae, or RSLs) were first identified by a University of Arizona student, Nepali undergraduate Lujendra Ojha, who was studying a crater over time. His professor, HiRISE

<< MRO and HiRISE continue to supply important science as well as pleasing images. When photos of this southern-hemisphere crater dune field photographed one Mars year apart were compared, features had shifted—an unexpected discovery.

SOLAR ECLIPSE >> Curiosity cameras can look up as well as across and down. The sequence of three images (above) shows Phobos, the largest moon of Mars, passing directly in front of the sun. The images were taken three seconds apart by the telephoto Mastcam on Sol 369.

IN ORBIT >> Cameras continued to monitor Mars remotely, catching this building storm (right) as it moves across the polar region.

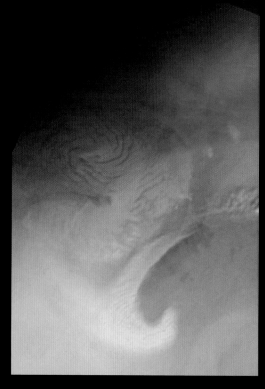

5:51 UTC (coordinated universal time) Orbiting cameras capture water, ice, and dust storms over Mars's north polar region.

principal investigator Alfred McEwen, took the research further, and together they published a paper in *Science* that broke the news. It was photogeology at a whole new level.

Since that first discovery, more than 25 sites have been photographed with streaks (and, in winter, without streaks), and some of the sites alone have more than 1,000 flows. No firm consensus exists yet that the streaks actually consist of water flowing just below the surface, but McEwen says that no other competing theories have emerged.

"We were actually taking images from these sites for four years before someone did the work to find that the slopes had these features in the warmer times and had none when it was colder," McEwen said. Those found so far have been largely in the mid-latitude and equatorial areas—far from the large ice deposits known to exist at the poles.

"It's quite possible these RSLs are all over the planet," says the still quite awed discoverer. If so, then the potential reservoir of water in ice form would be substantially greater than predicted, with global implications.

"We've actually looked at very little of Mars, and we keep on finding unexpected things that, frankly, we usually don't understand. It can't help but make you wonder: How many other surprising things are out there?"

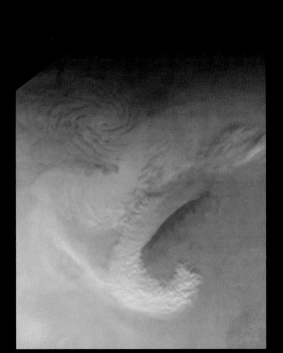

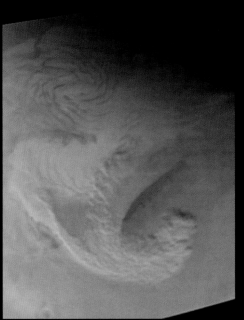

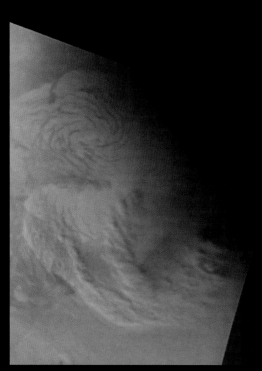

8:49 UTC The planet's polar ice cap is light in color, at the center of the image.

10:47 UTC Dust "curls" as it follows the largest of the storms.

12:44 UTC Mars Orbiter Camera shows the storm moving to the upper right.

PHOTOGRAPHY AT THE HEART OF SCIENCE

It was not that long ago that the field of Martian photogeology was not fully accepted. Too much room for misinterpretation, too little information about rock chemistry and minerals, never entirely definitive. After Curiosity, and its work in tandem with the orbiting satellites, those doubts have been largely put to rest.

It was Malin and Edgett's orbital discovery of layering at Mount Sharp that made Gale Crater appealing, and it was later HiRISE images and CRISM identification of clay and sulfate minerals that expanded that understanding and identified a landing ellipse for the rover. And then, with Curiosity on the ground, that process of ground-truthing the orbital data kicked in and quickly confirmed that the Peace Vallis fan had a geologically recorded record of flowing water.

It was MSL team members studying HiRISE images who also identified the light-toned unit of bedrock at what became known as Glenelg and later Yellowknife Bay, and helped lead Curiosity to a spot where great discoveries awaited.

Mars, it is now apparent, does have ancient secrets—many, many of them. The surface bleakness is still there, but it's looking more and more like an illusion.

CURIOSITY'S 17 CAMERAS

NAVCAMS Four Navcams look for the best path forward, all in black-and-white, and provide essential information to rover drivers. This shot shows the rim of Gale Crater in the distance.

RIGHT AND LEFT MASTCAMS Perched atop masts, these cameras create color photographs both long and narrow or broad and more shallow. These test photos study a rock outcrop south of Green River, Utah, Earth.

HAZCAMS Eight Hazcams help drivers avoid hazards, presenting images, also black-and-white, of boulders, inclines, outcrops, and filled or "ghost" craters. This image, taken with a Hazcam fish-eye lens, has been processed to create a level horizon.

Rover photography is essential for science, for driving, and for telling the story of Mars.

CHEMCAM Also producing black-and-white fish-eye images, this camera includes chemical analysis equipment and is used primarily for science. Mastcam or MAHLI color can be imported into a ChemCam photo.

MAHLI The Mars Hand Lens Imager camera, designed to serve as a stand-in for a geologist's field lens, provides very high-definition color images for both close in and far away.

MARDI One MARDI (Mars Descent Imager) camera took unprecedented images of the Curiosity landing from the exposed bottom of the rover. The camera is still taking pictures of the ground and rocks under the vehicle.

IN SEARCH OF ORGANICS

<< Yellow-orange carbonate minerals were found in the Martian meteorite ALH84001, which landed thousands of years ago in Antarctica and was discovered in 1984. The organic material was deemed evidence of past Martian life, but scientists have rejected many of the original claims.

CHAPTER 6

At the heart of the Curiosity mission, a quest for chemical compounds that are the building blocks of life

WORD WAS OUT and spreading: The SAM team had found something unexpected.

The Sample Analysis at Mars was the most complicated and sophisticated instrument on Curiosity, a chemistry lab the size of a microwave that can heat samples up to 1800°F (1000°C) and then track what gases are released in the oven.

SAM has many roles to play. But its primary and singular ability is to detect organic compounds—carbon-based molecules, usually bonded with hydrogen, that are the building blocks of life on Earth. While life on Mars or elsewhere could theoretically be organized around an element other than carbon, it's considered highly unlikely simply because carbon bonds so much better with other elements to form compounds useful to all forms of life.

As Curiosity science leaders often remind listeners, all life is made up of organic compounds, but not all organic compounds constitute life. A bacteria or a cat is made up of organic compounds, but so is an aspirin tablet, gasoline, and your favorite perfume.

So the detection of organic compounds on Mars would not necessarily mean that life is present or has ever been present there. However, a positive finding would definitely increase the odds that Mars was once home to biology. This would be especially true if Curiosity were to detect complex organics that are important to life on

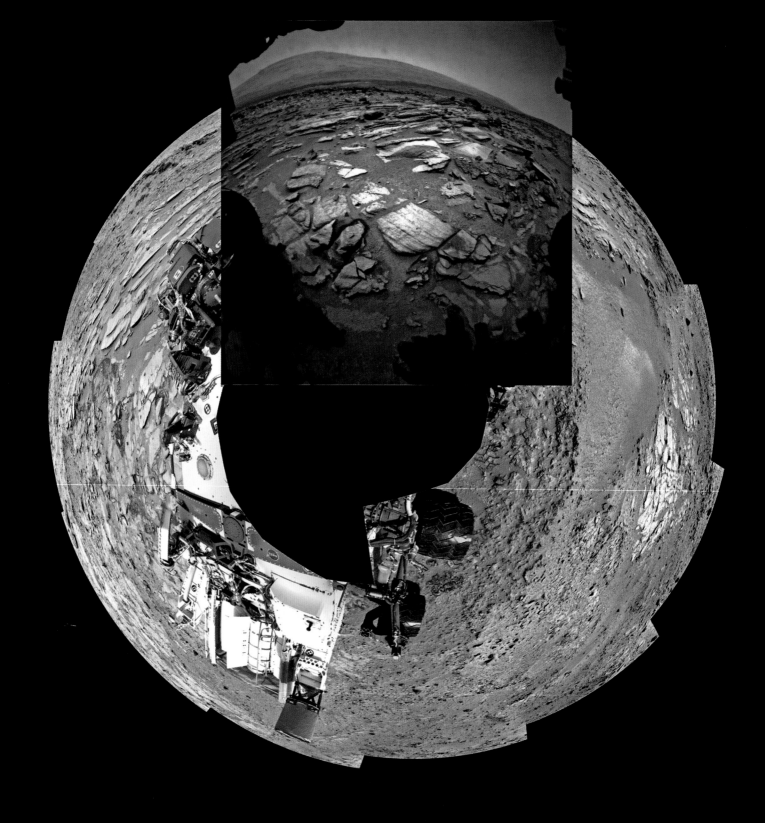

Earth, such as amino acids that are the building blocks of proteins and the lipids that make up cell walls. While these could be delivered to Mars via meteorites, they could also be biological in origin—what would be a clearly enormous and even historic discovery.

When tidings of a possible major discovery by SAM began circulating in mid-November (around Sol 100), Curiosity scientists and those following the rover's exploration sat up. Could it be organics?

Their discovery would represent not only a major scientific breakthrough but also resolve the decades-long mystery: Why weren't organics being found on Mars?

IMPLICATIONS OF ORGANICS

Scientists know for a fact that organic compounds rain down on Mars all the time from space in the form of meteorites, micrometeorites, and interplanetary dust. They are sure of this because the same happens on Earth, and no doubt other celestial bodies.

There is an endless supply of organic compounds floating through space, and some 30,000 tons of cosmic dust filled with organics falls on Earth each year. Back during what is called the "late heavy bombardment period" some four billion years ago, more than 1,000 times that much cosmic material (and thus organics) fell on both Earth and Mars.

So the question about organics on Mars was not whether they have been present but, rather, what was happening to seemingly destroy them on the surface? Or were scientists looking for the wrong things, and doing it in a way that would work on Earth but not Mars?

The possibility that organic compounds had finally been detected seemed all the more remarkable because SAM had just received its first samples of Martian material, and they were from a rather inauspicious site.

The first sample that went to SAM had been scooped on Sol 93—November 9, 2012—from a sandy drift given the name Rocknest. The site, about a quarter mile from Bradbury, had been selected for the first sample because the dusty sand seemed to be the right size for Curiosity's instruments, the site had some interesting geological features, and it was relatively flat (and so safe for a first attempt to gather material from Mars).

The rover was painstakingly maneuvered into position at Rocknest and, for the first time, the arm was extended to collect a small piece of Mars. This was no simple procedure: The arm is not only seven feet long, but also weighs 154 pounds and has a 65-pound turret on the end.

There were scientific questions to be answered, but Rocknest became the first sampling site largely for a different reason: housekeeping.

The location checked out as ideal for, in effect, cleaning out the tools used to deliver the samples. The sand would scour the several components of the delivery system, and that way hopefully remove all traces of Earth. JPL lead surface sample engineer Daniel Limonadi called it "rinsing and spitting."

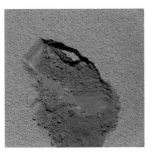 

FIRST SCOOP SITE The site (left) of a Curiosity scoop at Rocknest, where the robotic arm (right) first gathered Martian material.

So Rocknest was hardly a spot where discoveries were expected on the organics front: Sand, after all, is very poor at preserving organic compounds. Nonetheless, in early November 2012, several dozen Curiosity team members remaining at JPL gathered around a long table off the mission support room. Many more dialed in from around the nation and world to hear a report from SAM principal investigator Paul Mahaffy and his team of scientists.

Mahaffy gave all the appropriate and necessary caveats—that the team had worked with the sample for only a short time and that the absolute cleanliness of both the SAM instrument and the Curiosity delivery system remained uncertain. But an unmistakable excitement remained in his voice: SAM appeared to have found organic compounds. It also appeared to have found carbonate minerals and a chlorine salt of great interest to Mars scientists called perchlorate. But since identifying possible organic compounds was one of the central goals of the Curiosity mission, that finding was the headline.

Mahaffy, an expert in atmospheric chemistry at NASA's Goddard Space Flight Center, introduced the evidence and walked those present and on the phone lines through the workings of SAM. Then he turned to some of the scientists who had been analyzing the sample to give their presentations.

A dozen people spoke, but it was perhaps Danny Glavin, a high-energy and pioneering astrobiologist also from Goddard, who seemed to distill the findings and their importance. An origins-of-life-on-Earth researcher by training, he had gradually moved into the intertwined field of life beyond Earth.

With some 98 percent certainty, Glavin said, SAM had indeed identified organics—chloromethane (CH_3Cl) and dichloromethane (CH_2Cl_2). Simple molecules that combine carbon, chlorine, and hydrogen, they are present on Earth as well and have been used in products that strip paint, dissolve grease, and even decaffeinate coffee.

As Glavin described it, the detection could mean one of three things. Most exciting was the possibility that SAM had discovered Martian organic carbon. But it was equally plausible that trace amounts of organics from Earth had hitched a ride inside SAM and then were transformed to the chloromethanes through heating. Also, the CHIMRA (Collection and Handling for In-Situ Martian Rock Analysis) sample delivery may have carried trace amounts of organics from Earth and then inadvertently delivered them to SAM on Mars.

At that point, Glavin said, it was anyone's guess how organic compounds might have gotten into SAM. But he was no doubt speaking for much of the team when he said they would rigorously test all possibilities, but sure hoped the chloromethanes turned out to be Martian in origin.

ELUSIVE ORGANICS

The search for Mars organics has always long been a wild roller-coaster ride, and a car filled with SAM scientists was now making its way up what appeared to be a promising new hill.

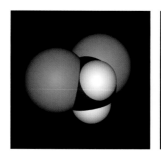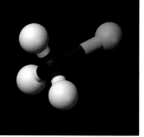

NATIVE OR IMPORTED? The organic compounds chloromethane and dichloromethane have been detected, but it remains unclear whether they were brought to Mars.

NASA scientist David McKay, lead scientist on ALH84001, displays the meteorite in a clean room at the Johnson Space Center. While McKay's findings have lost support over the years, the excitement surrounding his announcement ushered in a new age of interest in life beyond Earth.

>> Before the two Viking landers touched down on Mars in 1976, many space scientists had high hopes of finding signs of life, but the bleak landscapes photographed by both Viking craft (right) put Mars on the scientific back burner for two decades.

NASA's Phoenix spacecraft landed on the far northern plains of Mars in 2008 and discovered that water ice existed just below the surface and that the caustic compound perchlorate was present in the soil.

<< Scientists have long speculated about where the water on early Mars had gone, and Phoenix made clear that at least some of it remains below the surface as ice even today.

One of the scientists asked what the SAM data really meant. Dawn Sumner, a Curiosity long-term planner as well as mapper in chief and geologist, said with a Cheshire cat smile, "Mission success."

Or was it?

A discovery of Martian organics had the potential to be a turning point in understanding Mars, a game changer with endless scientific implications. And it would potentially move humanity a little closer to the possibility that we are not, or have not been in the past, alone in the universe.

On the other hand, the finding could become another in quite a long list of Mars discoveries turned controversies involving organics and possible signatures of ancient Martian life.

Scientists have been quite sure they had made such detections in the past, only to be met by skepticism, criticism, and ultimately broad rejection.

A case in point is the first successful mission to the surface of Mars, which featured the twin Viking spacecraft. Each landed on the planet in 1976 with the primary goal of testing for the presence of life. It was a bold if premature effort, and ended with substantial confusion.

One of the two instruments designed to search for life came up with findings that seemed to be positive—that microbial life was present. Many on the Viking team were skeptical and ultimately dismissed the finding after another instrument, designed exclusively to find those organic building blocks of life, found nothing organic at all. Without organics, the Viking science team concluded, how could there be life? Closing in on 40 years later, controversy continues to swirl around both findings—the initial "detection" of life, and the subsequent "non-detection" of organics.

While Viking was a great success in many ways, the confusion over organics and life helped put NASA's Mars program essentially on hold for years.

MARTIAN METEORITES

The next big Martian finding turned out to be equally ambiguous and contentious.

Many millions of years ago, a large meteorite hit Mars and kicked boulders into the atmosphere with enough force that some flew off into space. After some 16 million years of travel, a portion of the rock came into the environs of Earth and was pulled to the planet, landing in Antarctica.

Covered in snow and largely devoid of potentially contaminating surface life, Antarctica is the prime spot in the world for meteorite hunting. And in 1984, an American team found the rock, which at the time was not known to have come from Mars.

The rover, like the Viking landers before it, had
indeed found Martian organics.

But in 1996, a top scientist in the search for life beyond Earth, David McKay of NASA's
Johnson Space Center, authored a paper for the journal *Science* about the rock. His group
concluded not only that it had originated on Mars, but that it contained six unusual features—
including Martian organics—that together indicated it may well have once harbored Martian
organics and life.

The finding was hailed by NASA and highlighted by then President Bill Clinton, but other
scientists soon began to attack the science and conclusions of the McKay paper. They argued
that other processes and interpretations could explain the purported "biosignatures." McKay's
team fought back, and the man who led the team went to his death in 2013 convinced he had
it right. But the Mars and planetary-science community has largely dismissed the discovery,
except perhaps as a cautionary tale.

Still, the excitement around the possibility that life might once have existed on Mars led
NASA to greatly expand its emphasis on astrobiology—the search for life beyond Earth. And
since then, several researchers—including SAM team members Glavin and Andrew Steele of the
Carnegie Institution—have identified that organic material in Mars meteorites. Not ancient life,
but its carbon-based building blocks.

That growing NASA emphasis on astrobiology, however, did not include any formal search
for life or its component parts on Mars until Curiosity landed on the planet. Scientists had
been burned by high-profile claims that ultimately couldn't be substantiated, and realized they
needed a better understanding of how to look for Martian biosignatures and organics before
they mounted another astrobiology mission.

But ensuing NASA missions to Mars did provide insights, often unexpected, into what the
planet might be hiding. Perhaps the key discovery came from the 2008 Phoenix lander mission
to the northern polar region of the planet.

Its primary goal was to scrape a few inches into the soil and try to find water ice,
a task it accomplished before the frigid polar winter ended the mission. But in addition
to finding water ice, Phoenix identified the unexpected but persistent presence of the
salt perchlorate.

THE PRESENCE OF PERCHLORATE

The caustic compound, which is found on Earth and used here to treat thyroid problems as well
as in making rocket fuel, quickly became the subject of great interest.

Organic compounds do indeed constantly rain down onto Mars, just as they do on Earth.
Some even theorize that these organics from afar were essential to the genesis of life on early
Earth, and just possibly on early Mars, too. So why weren't they being found on Mars?

Meteorites, micrometeorites, and interplanetary dust all deliver organic compounds from space to Mars, Earth, and every other planetary or lunar surface. Although these organics from space are easily found on Earth, they have yet to be detected on Mars.

>> The very high profile ALH84001 Mars meteorite, on display at NASA's Johnson Space Center, has been dated as having initially broken from Mars bedrock about 3.9 billion years ago, during the time of a wetter Mars. Kicked off the planet by another meteorite impact some 15 million years ago, it remained in space until landing in Antarctica an estimated 13,000 years ago.

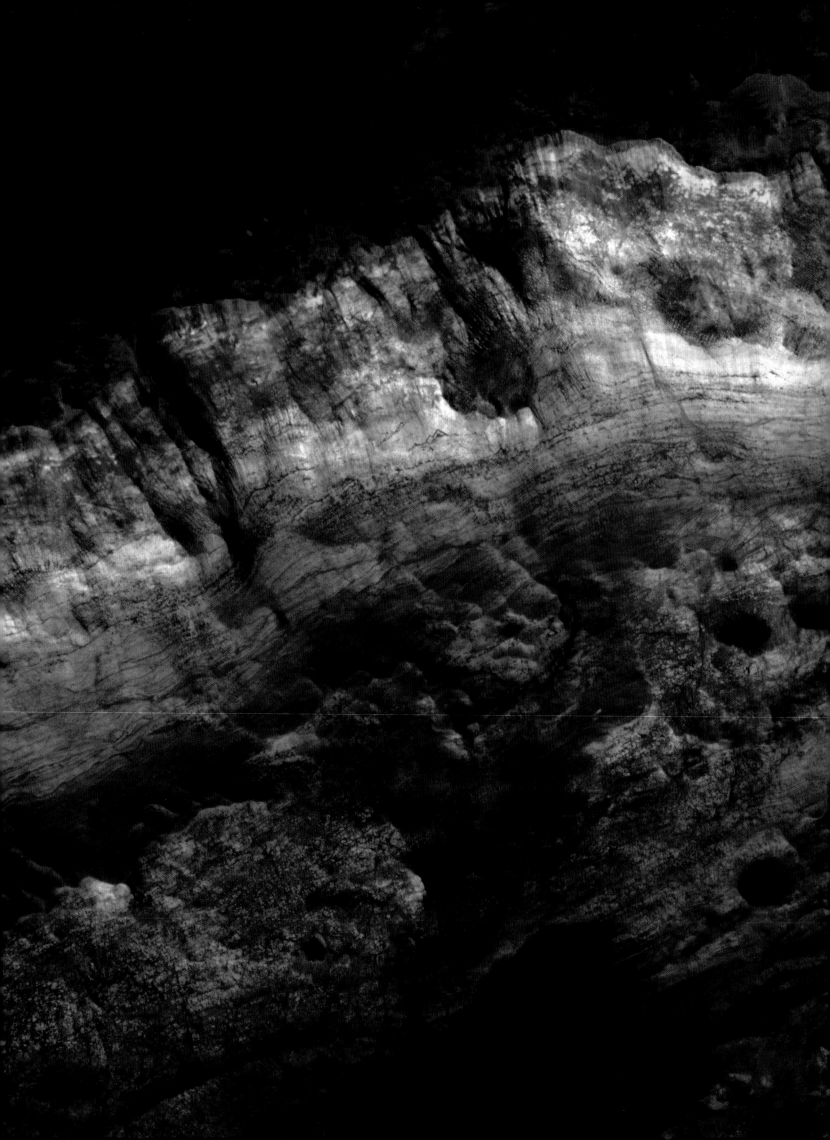

Perchlorates, which come in many different chemical varieties, just might hold the answer. They are found in some high deserts on Earth, and are known to modify organics when heated above about 300°C (572°F). The SAM oven, like the one on the Viking landers before it, did heat samples to that level and more, and apparently unintentionally destroyed or drastically changed any organics as it did.

NASA astrobiologist Chris McKay and early Earth and Mars specialist Rafael Navarro-Gonzalez, both members of the SAM science team, proved that in 2012. They heated some common organic compounds with a perchlorate, and the only organic compounds that they found afterward were dichloromethane and chloromethane. These are the organics found by Viking and now Curiosity after heating their Martian soil samples.

The Mars organics story was beginning to come full circle. That's because, little known to the public, the early Viking finding of "no organics" actually came with an asterisk.

As with Curiosity, samples of Martian soil collected by Viking were heated well above 200°C and produced gases that were identified as chloromethane and dichloromethane—which are, indeed, organics. At the time, however, they were dismissed as contaminants from Earth, leftovers from the effort to scrub the Viking lander clean, or perhaps from the fuel used during landing.

But now the discovery by SAM and Curiosity of those same two organics in a sample that also contained Martian perchlorate was perhaps telling an important story—that the rover, like the Viking landers before it, had indeed found Martian organics. In the presence of the perchlorate, however, they had been turned into compounds that could be easily dismissed as contamination from Earth. So could the perchlorate begin to explain why it has been so hard to find organics on Mars?

RUMORS OF LIFE GREATLY EXAGGERATED

It's no wonder that the SAM finding of organic compounds generated great excitement on the team; the detection involved a potentially paradigm-shifting discovery. But it remained a finding that was complex, far from crystal clear, and—as Curiosity leaders soon found out—slippery and very highly charged as well.

An inherent risk associated with talking about organics on Mars is that so many people assume that if something is organic, then it's alive. Or was once alive. It's a perhaps understandable error because the word *organic* arose from the ancient belief that certain compounds could be produced only by the action of a life force in organisms. But that's not the case.

Organics can be part of living organisms, or they can have nothing to do with life and biology at all. However many times Curiosity scientists explain that their search for organics is focused on finding carbon-based compounds that can be used by living organisms—and consequently is not a search for life itself—some will nonetheless conflate the two.

And that's exactly what happened when news began to leak out about the possible discovery of organics: The Web was alive with rumors that Curiosity had found Martian life, or at least that the rover had found something very big. When project scientist John Grotzinger, in a

<< The sedimentary sublayers of Mars are apparent in this crater wall in the area of the Mawrth Vallis channel. The bedrock outside of the crater is relatively dark, but this interior wall, exposed by a meteorite, reveals lighter and layered bedrock of different colors. A few dark patches on the crater walls are interpreted as dark sand in pits.

moment of excitement, told a radio reporter that the quality and range of information coming would make the mission and its discoveries one "for the history books," the quote and the story went viral.

"If it's going in the history books, organic material is what I expect," planetary scientist Peter Smith from the University of Arizona's Lunar and Planetary Laboratory told *Wired* online. "NASA may have discovered signs of life on Mars—but it's not telling. Yet," wrote the *Huffington Post*.

It was an awkward time as Grotzinger, and JPL, sought to correct the rumors that Martian organics, or even Martian life, had been found. As Grotzinger later explained, he had been misunderstood and was talking about the quality of data, not a particular finding.

"When this data first came in, and then was confirmed in a second sample, we did have a hooting and hollering moment," he said. "The enthusiasm we had was perhaps misunderstood. We're doing science at the pace of science, but news travels at a different speed."

SAM'S UNINVITED GUEST

Meanwhile, the SAM team was having a crisis of its own. As the new results were being analyzed, the team uncovered a most disconcerting problem: Some organic material used in the "wet chemistry" section of the instrument had apparently leaked out of its container.

SAM works primarily by heating Martian samples in its oven and analyzing the gases that come off. It brought up 59 cups to hold soil and rock samples that would be sent into the oven. But it also carried some cups where wet chemistry could be performed that did not involve heating rocks into gases. The wet chemistry would be used if it looked as though complex organics—which would be destroyed in the ovens—might be found.

Those sealed cups hold an organic fluid called *N*-methyl-*N-tert*-butyldimethylsilyl-trifluoroacetamide, or MTBSTFA. It is used to react with potentially complex Martian organics and form gases at low temperatures that can then be analyzed and identified by SAM. But if MTBSTFA is introduced directly into an oven—as opposed to the low-temperature cups—it will show up after heating as an organic compound.

And as the SAM team was analyzing their organic results, it became clear that an earthly contaminant had indeed mixed with the sample, and that contaminant was none other than MTBSTFA. Somehow one of the sealed cups had been breached and some liquid had gotten loose.

So the SAM team had a big problem: At least some of the organic compounds found in the sample were clearly produced when the MTBSTFA (a mouthful that came to be called "Mr. M" or "Mush") was heated in the SAM oven in the presence of perchlorate. A discovery of organics on Mars had again quickly turned uncertain.

As Mahaffy later explained, Curiosity and SAM had certainly found something important at the Rocknest sand pile; tests had indeed identified organics. Despite the contamination issue, it remained possible that some of the chloromethane and dichloromethane were in fact a by-product of the heating of Martian organics in the presence of perchlorates, as opposed to earthly contaminants.

"SAM is an amazing instrument, but if we had these samples in our lab here, we would have figured all this out in a week."

Glavin began as a terrestrial origins-of-life scientist and over time migrated into astrobiology and the search for both organic compounds and life beyond Earth. He has extensively studied meteorites and, with colleagues, was the first to find complex amino acids in a (non-Martian) meteorite that fell into the Nubian Desert of Sudan in 2008. Two years before, he and colleagues at Goddard identified nitrogen-rich compounds in the comet Wild 2, based on information sent back by the NASA's Stardust space-craft. His earlier origins work was done as part of a NASA fellow-ship in the lab of noted prebiotic-chemistry expert Jeffrey Bada at the Scripps Institution of Oceanography in San Diego.

Daniel Glavin, Goddard Space Flight Center

An astrobiologist with a planetary sciences and physics background, Glavin has played a key role in the search for organic compounds using SAM, the instrument that can analyze powdered rock samples (below).

"We really consider this a terrific milestone," Mahaffy said of the SAM findings. Nonetheless, caution prevailed, and he wanted it known that "SAM has no definitive detection to report of Martian organic compounds with these first sets of experiments. The reason we're saying we have no definitive detection of Martian organics is that we have to be very careful to make sure both the carbon and the chlorine are coming from Mars."

Project scientist John Grotzinger, no doubt mindful of the difficult and controversial history of the search for Mars organics and life, summed up the situation: "There's not going to be one single moment where we all stand up and, on the basis of a single measurement, have an hallelujah moment."

THE ROLLER COASTER

Finding organics at the sand pile of Rocknest always was a stretch: The sandy soil is not good at preserving whatever organics might once have been there.

But finding them in clay or sulfate minerals—which are formed in water and tend to preserve organic compounds better than most geological settings—was considered far more plausible. And so the SAM team was eager to test its first rock (as opposed to sand) samples, a complex and time-consuming process that would not occur for 40 more sols.

To prepare for those drilled, powdered samples, the SAM team looked for work-arounds to deal with the problems of the MTBSTFA contamination and, even more difficult, the perchlorates.

Jennifer Eigenbrode, a geologist and bio-geochemist at the Goddard Space Flight Center and a member of the SAM science team, specializes in detecting Martian organics and determining the effects of radiation on organic compounds. Here she injects a chemical into a rock sample to test whether it preserves the rock's molecular structure.

Chris McKay of NASA Ames—a veteran astrobiologist who has worked in almost every extreme place on Earth you could image—had been studying the perchlorate issue for several years. At first he was excited when SAM identified the salts in Rocknest; it meant that they were most likely common around the planet, which could have a broad range of implications.

The various forms of perchlorate lower the freezing point dramatically (allowing water to remain a liquid brine down to -58°F [-50°C], for instance) and can also be used as a food source by some bacteria and other microbes on Earth. So perchlorates, which on Earth are generally seen as harmful, just might be an important positive in terms of possible life on Mars and other planets.

But in his cluttered office in Ames, located at the decommissioned Moffett Naval Air Station near San Jose, McKay said "there's nothing good about perchlorates when it comes to looking for organics."

When heated in the presence of organics, large amounts of oxygen are released, and those chloromethane compounds are formed and released as well. For McKay and his SAM colleagues, this "perchlorate wall" makes it extremely difficult to identify any Martian organics, even if they're present in a sample.

"The bad news here is that the perchlorates seem to be destroying any organic material delivered to the instrument, and may be doing that on the surface before we even get it," McKay said. "The way I explained it to some students recently was like this: We're looking at the burned ashes of a house and trying to figure out what it was like living there before the fire. Not a great way of doing business."

But the search nonetheless continued, and even with some promise. At Goddard, Glavin and his colleagues scrubbed the data they were getting from Rocknest and the two later sample sites—called John Klein and Cumberland—and found concentrations of organics not easily explained by the MTBSTFA contamination and the effects of perchlorate. Signatures of at least two other carbon compounds (chlorinated methanes and a chlorinated benzene) also showed up.

"The whole process has been a roller coaster," Glavin said. "One day there's a lot of excitement, then the next maybe something disappointing. But many of us believe we are seeing something Martian in there, even with the perchlorates and the MTBSTFA background. We see different abundances of the chloromethanes at different sites. I think the organics story we can tell is getting more clear."

Long hours of high-pressure lab work had given Glavin and his colleagues at Goddard insights into how to work around the perchlorates and the "Mr. M" background.

"We've knocked the MTBSTFA way down and some organic signatures remain," Glavin said. "We're definitely not home, but it's looking increasingly like we have organic material in our sample that isn't internal contamination, that we did not bring with us."

In other words, possible Martian organics.

BEFORE AND AFTER Matching photos show a patch of soil, named "Beechy," where ChemCam shot small holes to test for chemical variability.

GOING AFTER MORE

Around the same time (nearly Sol 400), SAM colleague Jennifer Eigenbrode was returning her team to the organics roller coaster with some potentially dramatic new findings.

Eigenbrode, a biogeochemist with a background in studying the early Earth, also works at Goddard and alongside Glavin, Mahaffy, and the largest instrument team on the Curiosity mission. She had started examining in finer and finer detail some particular results from the SAM oven when it was burning at some of its highest levels. She thought she saw some small, broad elevations on the SAM graphs, but she couldn't tell if they were artifacts or the signature of something important.

One of her colleagues, Doug Ming of NASA's Johnson Space Center, had been reminding his colleagues to always "expect the unexpected on Mars," and it would have been quite unexpected to find something important in the results she was combing through. But with his words in mind, combined with a broad knowledge of how organics react to the kind of conditions present on Mars, she was drawn into a deep dive. First she did it all herself, and then she brought in a software expert to help automate and speed up some of the processing.

"Basically, we worked nonstop for three weeks—weekdays and weekends, with very little sleep. The results were getting more and more interesting, and that kept us going."

This was not data from the Rocknest scoop sample she was looking at, but SAM data from Curiosity's first and second drill sites. And with the readout from the last sample—which was a triple dose of pulverized Martian rock—she was ready to present to her SAM colleagues. It was quite possible, she told them, that fragments of large organic molecules were showing up at the high-temperature portion of the experiment, that the small rises were not background noise.

The way to test that conclusion was to run a SAM blank—a control that would heat one of the empty cups that holds the sample to see if those small rises on the SAM readout graph were present. If the small peaks showed up again, then the signal Eigenbrode was seeing was noise, an artifact of the SAM instrument and processing.

But if they were absent, then that meant the rises she had seen had come from the sample. And if they came from the sample, they had a good chance of being signatures of Martian organic compounds.

The blank test was run on Sol 409.

The results: The blank showed no high-temperature peaks. That meant SAM had quite likely detected something in the earlier sample itself that came from Mars rather than from the instrument.

Eigenbrode was ecstatic, but not for long. Scientists quickly move to the next phase of trying to knock down their own findings. Others on the SAM team also joined the chase—trying to replicate the results, taking them to the Goddard lab to better understand the chemistry, looking for ways to understand what Mars was telling them.

Months of work, for sure, but now the Martian organics roller coaster was climbing up again.

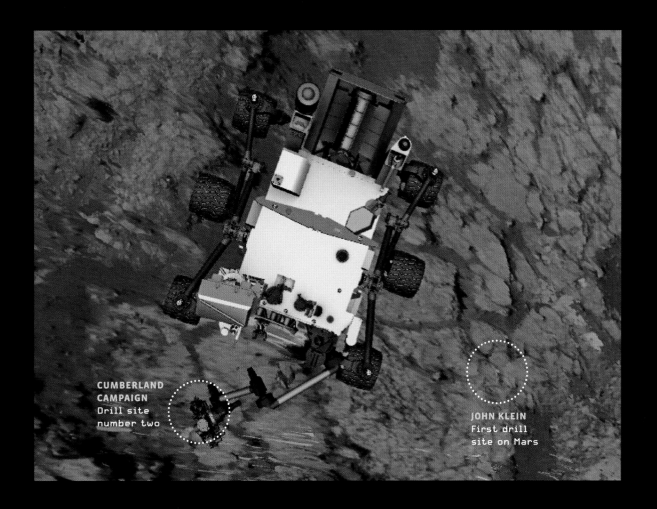

CUMBERLAND
CAMPAIGN
Drill site
number two

JOHN KLEIN
First drill
site on Mars

CHEMICAL ANALYSIS >> The rover approaches the flat, fractured Cumberland drill site in Yellowknife Bay in this overhead view (above) generated by software used by the rover drivers. Analysis of the powder drilled at Cumberland showed some organic compounds, perhaps Martian in origin. Laser beams bounce between mirrors in the measurement chamber of the Tunable Laser Spectrometer in SAM (below) during a demonstration of the instrument. On Mars, the laser beams used are in the infrared range, and so would not be visible to our eyes.

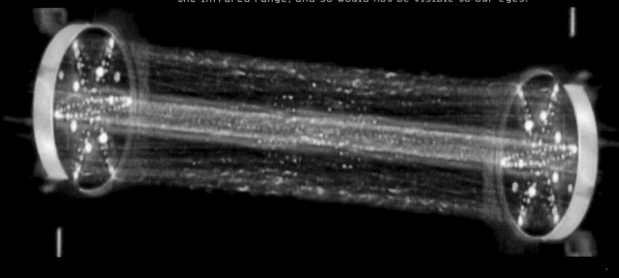

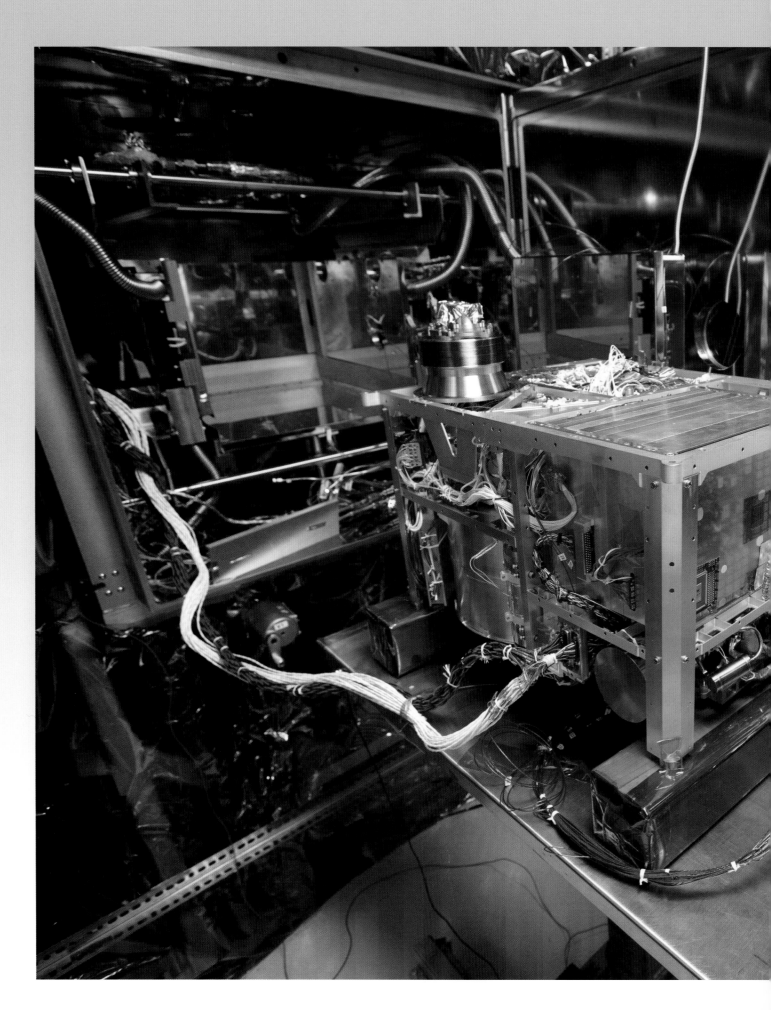

An exact replica of the SAM instrument on Curiosity is kept in a Mars-like environment at NASA's Goddard Space Flight Center in Maryland. Only with great care is it taken out for upgrades or maintenance, as it was in early 2013.

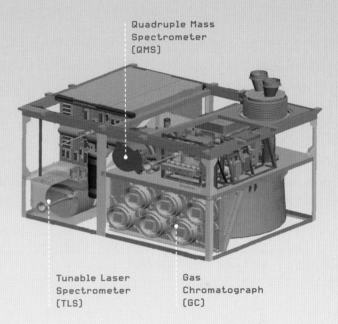

Quadruple Mass Spectrometer (QMS)

Tunable Laser Spectrometer (TLS)

Gas Chromatograph (GC)

SAM is often described as the most important and complex piece of equipment ever to land on another planet. The gold-covered box holds two tiny cylinder ovens that can vaporize Mars's rocks and soil at temperatures up to 1800°F (1000°C). Three spectrometers then identify and analyze the gases produced by the ovens, as well as those collected from the Martian atmosphere. Some six miles (nine kilometers) of electrical wire connect these and many other parts. SAM's task constitutes a primary aim of Curiosity's mission: investigating whether Mars preserves the chemical ingredients needed for life, including organic carbon.

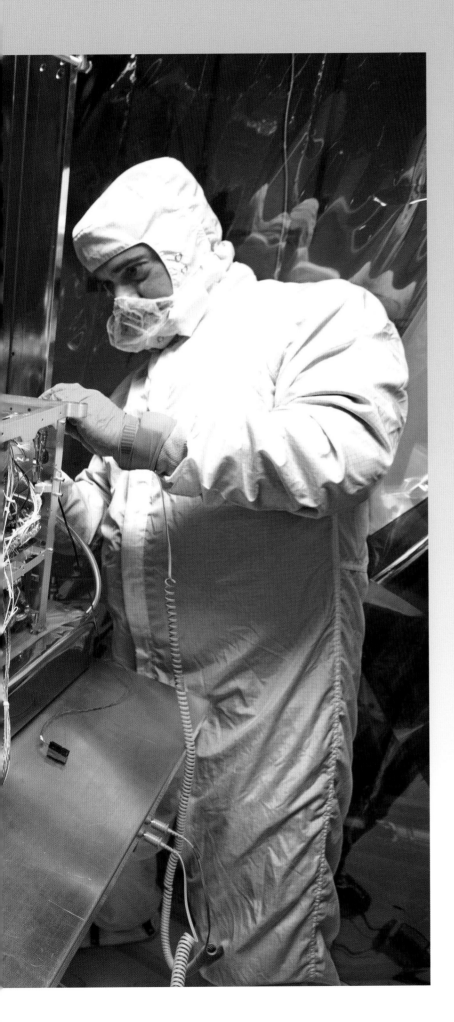

A HABITABLE PLACE?

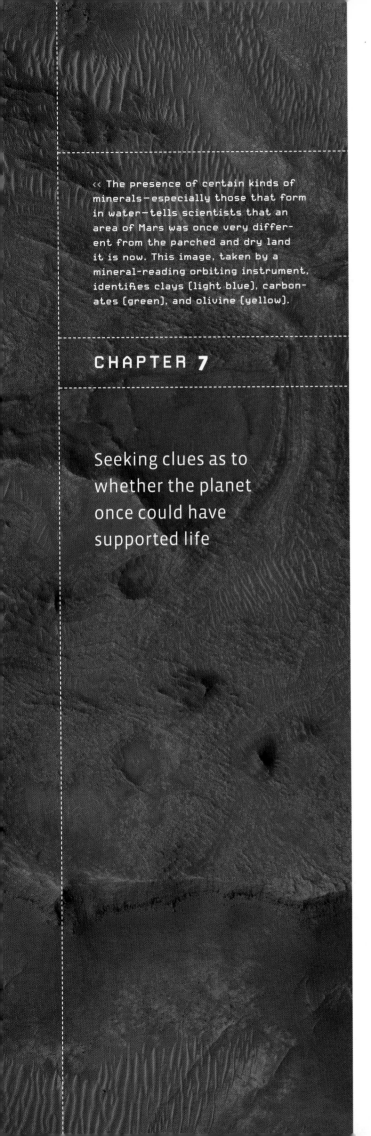

CHAPTER 7

Seeking clues as to whether the planet once could have supported life

SCIENTIFICALLY SPEAKING, the journey into the depression near Glenelg did not begin auspiciously.

Intriguing surprises cropped up along the way, but most of the early results from ChemCam, and then the APXS, identified volcanic rocks along the pathway that showed no signs of having been altered by water. The Rocknest outcrop was probably volcanic, too, and so were boulders and bedrock in the area. A minority theory that the depression was a lava bed always inhospitable to life—rather than a pond or lake where primitive life-forms might have originated—began to sound increasingly plausible.

For Bill Dietrich and some others who had argued so hard for the detour, those early weeks of the descent into the bowl were disappointing and disturbing. In fact, Dietrich says they were some of the worst days of his adult life.

That he had argued so vociferously for what was looking more and more like a costly misstep was quite possibly giving him ulcers, and even making him wonder about geomorphology, the scientific discipline that had guided his career.

Neither was it an easy time for John Grotzinger, the project scientist, a star in the world of Mars geology and science, and the most important advocate for the Glenelg detour. His interest in the area had come from a somewhat different geological perspective than that of Dietrich and

The science wasn't definitive, but many see high
thermal inertia as the sign of a watery past.

others eager to identify a lake; even before landing, Grotzinger had been intrigued by the triple contact point of rock units that he and others identified via satellite imaging.

But an equally strong magnet was that known presence of rocks with high thermal inertia, that ability to keep ground temperatures from having such dramatic temperature swings. The science wasn't definitive, but many see high thermal inertia as the sign of a watery past. So for Grotzinger, too, the detour involved a rather high stakes search for signs of something resembling an ancient lake.

Grotzinger has written well-received papers by the score, but he is also known for his emotional intelligence—his instinct to encourage debate and inclusiveness, coupled with a willingness to make hard decisions when necessary. Some said that as Curiosity project scientist, he was charged with herding cats, since he had to lead 367 other scientists who had strong opinions of their own.

By most accounts the man and the task were made for each other, even if they were certainly tested on the way to Glenelg. The quick look-see that he and others had proposed, and that began on Sol 22, was starting to look like a major investment of time and opportunity.

It was around Sol 120 of the traverse, when the rover had traveled a bit over a half mile, that the landscape began to change significantly as Curiosity headed farther down a gentle slope.

Outcrops with clearly layered features—and given the names Point Lake and Shaler—were spotted via Mastcam, and the rocks took on more of a sedimentary look. While Mars is a primarily volcanic planet, the hopes for habitability rested with locations where dislodged sediment carried by water, wind, mudslides, or ice had collected and been transformed into rock.

The lowest point they would reach was given the name Yellowknife Bay, a reference to the iconic area in Canada's Northwest Territories where geologists head in search of some of the oldest rocks on Earth.

On Sol 120, the rover left Shaler and entered what, from orbit, appeared to be the white-toned area where the ground temperatures were less extreme. This was a big moment for Curiosity: Would its own weather station similarly find a new ground-temperature pattern as the rover entered the area of high thermal inertia?

The answer came quickly and was positive. The temperature of the rocks they had recently driven over dropped down to -120°F (-84°C) at night and rose up to 32°F (0°C) and above during the day. But on the rocks of the white-toned area, the nighttime ground

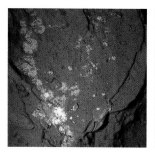 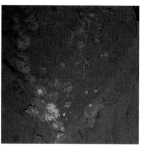

DAY AND NIGHT LED lights allow the MAHLI camera to take photos at night. These two images of the same rock were taken at Yellowknife Bay.

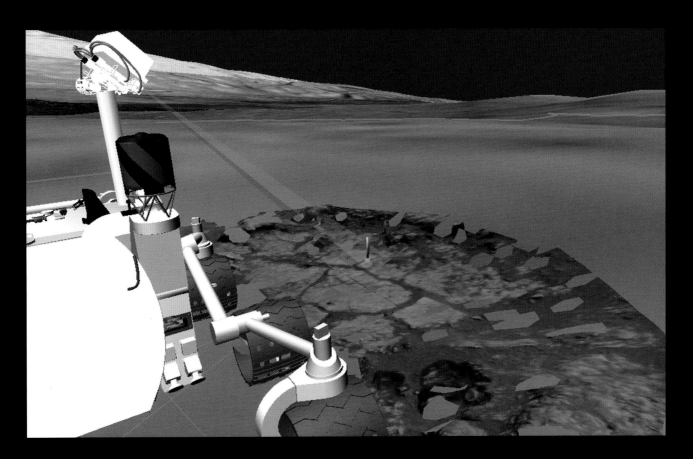

READY, AIM, VISUALIZE >> To increase the odds that a Mastcam photograph will be the one desired, first a visualization is made of the rover and the science target. This visualization is clearly from Yellowknife because of the flat, platelike, fractured rocks.

STUDYING YELLOWKNIFE BAY

137°27'E

MAP KEY

* Landing site
342□ Way point and mission day (sol number)
— Traverse path

John Klein and Cumberland

YELLOWKNIFE BAY

130

Point Lake

297

4°35'22.5"S

Rocknest

125

Bathurst

124

111

123

From Bradbury Landing

53   59

324   307

50

329   327

49

GLENELG   317   Shaler

331

333

Continued Traverse

METERS
0   50
0   200
FEET

DETOUR TO YELLOWKNIFE

The Curiosity team took a major detour from its journey to Mount Sharp by heading to a depression near the landing site it named Yellowknife Bay. The area proved to be a scientific gold mine that the rover explored and analyzed for months.

"Everything we measure on Mars has information embedded in it. Our job is to understand it in the fullest Martian context possible."

Grotzinger initially embraced geology as a young man after realizing it allowed him to work in his other favorite fields—chemistry, biology, and physics—and to pursue them all outdoors. As project scientist for Curiosity, that comfort and pleasure in mixing disciplines has been an essential impulse as he leads the effort to piece together the wide-ranging information collected by the rover instruments. In addition to his role at JPL with Curiosity, he is the Fletcher Jones Professor of Geology at Caltech. Among his many areas of interest is geobiology, the study of the back-and-forth interaction between organisms and the nonliving world. He has been studying Mars hands-on for some time, having been a scientist for the Opportunity and Spirit rovers.

John Grotzinger, California Institute of Technology

A geologist by training, Grotzinger is Curiosity project scientist for the mission, in charge of more than 350 scientists. He seeks consensus about where the rover should go, such as making the detour to Yellowknife Bay (below).

temperatures seldom got below -65°F (-54°C), and during the day they seldom rose above 0°C. Something very interesting had changed.

What's more, the rocks looked flatter and softer, more eroded, and some even came in the kind of polygon shapes associated with ground freezing and thawing in the presence of water.

THE EUREKA MOMENT

It was ChemCam that provided the eureka moment: Using its laser blaster, it kicked up a gaseous plasma that the instrument then identified as containing sulfates—signature minerals that form only in water. It was beginning to look as if Yellowknife really did go through a time that was wetter and probably warmer.

As the rover inched into the Yellowknife bowl, a gloriously sedimentary world came into view.

Many of the rocks were very fine-grained, had big fracture lines, and appeared to be "mudstone," a type of soft rock that requires a watery past. Pebbles and fine-grained sands were common, and some of the rocks were definitely conglomerates, formed in the presence of water. Scientists spoke of similarities to Death Valley, to Antarctica, to southern Australia.

Then ChemCam and Mastcam began to identify veins of white and calcium-based material in the rocks, as well as bumps, known as concretions, that are small rounded concentrations of minerals formed in water.

These discoveries fit well with the theory that the Peace Vallis river or stream had spread as far down as Yellowknife, and had done so at different periods of time. The white veins that filled cracks in the mudstone were perhaps the best clue. They told the story of sedimentary rocks initially formed in water, then later covered again with water carrying the compounds that could lay down the gypsum-like veins.

This was a hidden Mars unlike anything witnessed during decades of exploration, Mars science at its absolutely most exciting.

But to get beyond the watery logic presented by the cameras and the instruments that primarily studied the rock surfaces, Curiosity needed to get some rock samples into the chemistry labs of SAM and CheMin. And that meant drilling down, getting several inches below the surface. The grayish blue insides of the rock broken by a wheel of Curiosity whetted the scientific appetite; the team was anticipating something similar to be brought up by the drill.

Given what was before them, the decision to take the equipment out of storage and begin the first drilling campaign ever on Mars was an easy one to make. Extremely difficult to actually perform, but the obvious next step for Curiosity.

Curiosity drilled this hole the depth of a golf tee and the width of a Sharpie to collect powdered rock for analysis. The line of small holes within the hole were made by ChemCam laser shots; the white streaks show deposits of the mineral calcium sulfate. The detail within the hole was captured by the MAHLI camera.

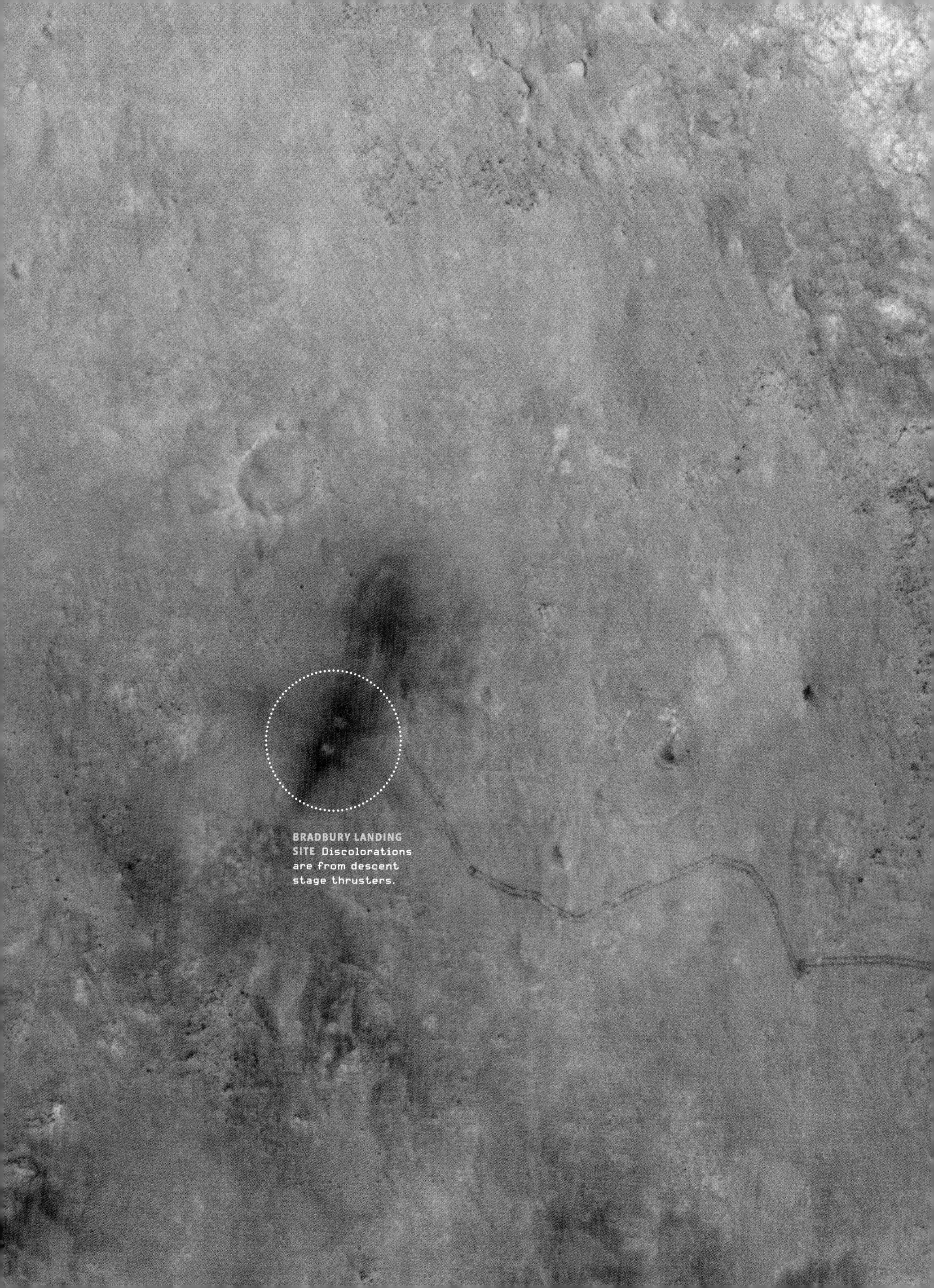

BRADBURY LANDING
SITE Discolorations
are from descent
stage thrusters.

YELLOWKNIFE BAY
First site declared
habitable on Mars

DIGGING DEEPER THAN EVER

There's a good reason why no tool sent to Mars had ever dug more than a minuscule distance below the surface: It's complicated, potentially risky to the health of the entire rover, and costly to do.

The technical person selected to lead the design, development, testing, and delivery of a Mars drill was Louise Jandura, a mechanical engineer from blue-collar Clifton, New Jersey. Her father ran a trucking and excavating business, and she had never traveled more than three hours from her home before she was enticed to the Massachusetts Institute of Technology.

She came to JPL in 1989 and gradually worked her way up the ladder of responsibilities. Leading a team that conceptualized, designed, tested, and now helped to operate the system has been her task for eight years.

But that was only part of the job. She and her colleagues also had to create a system for capturing the powder ground up by the drill, inserting it into a delivery mechanism at the end of Curiosity's seven-foot arm, shaking out powder grains that are too large or small, and finally dropping the sample into a small hole on the deck of Curiosity that opens to receive the precious delivery of Martian rock.

But the great innovation was the drill. In the years getting ready for this first piercing of a Mars rock, Jandura says, "we made eight drills and bored more than twelve hundred holes in twenty types of rock on Earth."

All the men and women chosen to lead a crucial activity have their moments of truth, and Jandura's was drill day on Sol 182 (February 6, 2013). For Adam Steltzner and the EDL team it was the landing; for Jennifer Trosper and the "brain transplant" team, it was when the computer software was switched; for all the instrument principal investigators, it was when the first experiments were tried and results either came in or didn't.

The location selected for the first drill was on a flat stretch of Yellowknife, with fractured plates of mudstone that

<< Curiosity's path from Bradbury Landing to the outskirts of Yellowknife Bay is visible in the Mars dust. The distance was less than half a mile (700 meters), but it took more than five months to get to the endpoint.

The team that developed and operated the first ever drill on Mars has been led by Louise Jandura, upper right. Before beginning its first dig at the site named John Klein, the team tried out eight different variations.

suggested a long-ago lake bed. The site was named after recently deceased deputy project manager John Klein, and the campaign was expected to take a week.

Instead, it took 24 sols from the first checkout for pre-loading the drill (on Sol 170) until the first delivery of rock sample to the CheMin lab.

Step one was to maneuver the arm close to the ground so the drill prongs with their sensors could actually find the surface. It's a tricky maneuver.

The drill is part of a 66-pound (30-kilogram) turret that sits at the tip of a 150-pound (70-kilogram) arm with five joints. They're called five "degrees of freedom," and they consist of two joints at the shoulders, two at the wrists, and one at the elbow. They have to move in concert: Choreography had been practiced many times at JPL's test bed.

The first contact was made successfully. The drill prongs were touching Mars.

The target was one inch in diameter, and the drill came in well within that. Then it was time to push down, always rotating and hammering. The drill made a small divot in what turned out to be soft stone.

If rover computers detect a problem—if the approach angles are too sharp or the rock is too hard—fault protections will stop the whole process. And several times they did: The checkout command to hammer, a test mini-drill, and the transfer of the collected sample all had to be carried out more than once.

The first full drill attempt was made on Sol 182. Digging down at what was considered an impressive 8 or 9 millimeters per minute, the drill rammed its way to 6.5 centimeters (2.5 inches)—its limit. The hole was the height of a golf tee and the width of a finger. The powder pulled out was later estimated to be about one tablespoon of never-before-encountered sample from inside a Martian rock.

The actual drilling was finished in ten minutes. The drill bit was then slowly pulled out, and the powder was held in an appendage of that assembly.

It took 11 days to get a photo confirming that the sample had indeed been successfully collected and transferred to the sample processing hardware, which has the ungainly name of the Collection and Handling for In-Situ Martian Rock Analysis tool, or CHIMRA.

Over the next few days, CHIMRA (pronounced "chimera") shook and rotated the powder to prepare it for testing and to clean the drill. The sample was brought to the rover in a kind of reverse robotic arm maneuver. It was then placed next to the collection box that receives the goods and dropped into the rover.

Much of the action was captured by the MAHLI camera, creating a visual record of the first time that humans cut into a sedimentary piece of Mars bedrock and delivered the samples to chemistry labs inside the rover.

DRILLING DOWN >> The first Mars drill was carried out on flat land and into soft mudstone rock (below). The selection was intentional: Despite all the practice, managers wanted to give the drill an easy first time out. But the mechanism is designed with protections: Should the drill become stuck in a hard rock, it can cut free from its metal bit (above, left) and replace it with a spare on the outside of the rover. Before beginning the first dig on Mars, the team bored more than 1,200 practice holes on Earth in 20 types of rock (above, center). Researchers visualize every drill event before they command the rover to perform it (above, right).

We now know the drill was a success. But Jandura was in the Mission Control room for the daily downlink during the whole campaign—waiting, she says, like an athlete before a big game. With ever changing Mars time, those downlink times were ever changing, too. It was 5 a.m. when the biggest downlink arrived, the one showing the first image of powdered and collected Mars rock, safe and sound in a small collecting bin.

But when that image was flashed, Jandura and her drill colleagues whooped it up. Some were with her—including her key teammate, the "cognizant engineer," or "coggie," Avi Okon—while others were at home and on phones or chat lines because of the time.

Jandura recalls that early morning downlink with a pleasure that wouldn't be matched by the many later drills. "This first drill could have been just an engineering exercise. I can't describe how immensely cool it is that we got to use it on something they care about a great deal on our first shot."

It was a great time for JPL engineering, and the science news would be equally grand. Indeed, it would be revelatory. The powder dug up from beneath the cover of the red planet wasn't red at all, had none of the caustic iron oxide rust that defines the atmosphere and surface.

Instead, the powder was grayish blue.

RETHINKING THE RED IN RED PLANET

The first color images of the drill sample were, not surprisingly, received with jubilation. The red planet was really not red at all, but rather was wearing (and breathing) a covering of rouge. In the world of geology, red is associated with iron, rust, acids, and environments tough on organic or living things, while grayish blue is the color of far friendlier conditions.

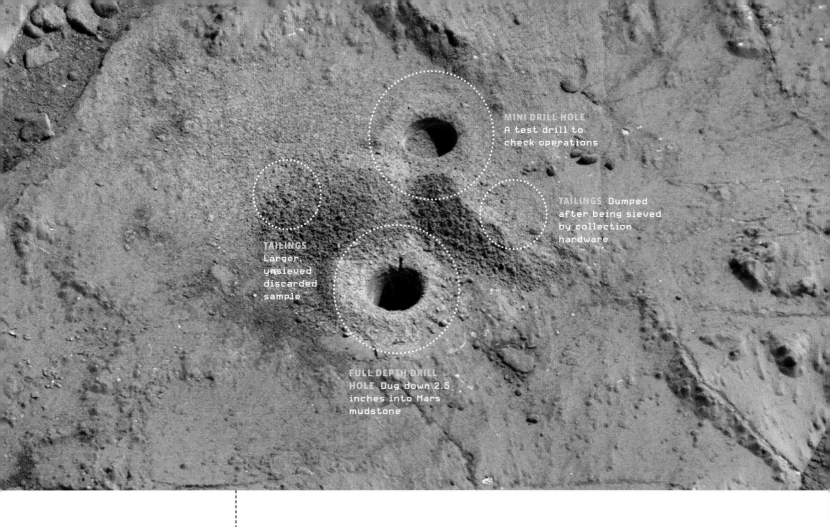

MINI DRILL HOLE A test drill to check operations

TAILINGS Dumped after being sieved by collection hardware

TAILINGS Larger, unsieved discarded sample

FULL DEPTH DRILL HOLE Dug down 2.5 inches into Mars mudstone

It took several weeks, but when the samples were delivered to SAM and CheMin and then analyzed and reanalyzed, the results that came back were all the Curiosity scientists could have hoped for.

The premier finding was the presence of clay minerals called smectites, and the concentration was as high as 20 percent of the sample. Clays are phyllosilicate minerals laid out in flat, hexagonal sheets, rather like mica. Their structure is based on combinations of silicon and oxygen, but other elements are often present and produce variation. But for all of them, the process through which they become rock minerals requires substantial amounts of water.

On Earth, clays are generally neutral in the scale of acid to alkaline, and consequently great for finding fossils of all kinds. The John Klein clays were similarly formed in a setting that was neither particularly acidic nor alkaline, the testing showed. While microbes have been discovered on Earth that live in highly acidic or basic settings—they're called extremophiles—early-life researchers generally hold that Earth's first microbes formed in the kind of neutral pH found at Yellowknife. And clays, unlike more acidic minerals, are known to preserve the carbon-based organic compounds being so aggressively pursued by the Curiosity team.

Adding to the excitement was the detection of the mineral olivine, a substance formed only in the Martian mantle, in volcanoes, and then on the surface after spewing out.

Olivine on the surface is known to be a common precursor to clay, and is formed to some extent by its decomposition. That olivine was mixed with the clay added to the possibility that the clay was formed where it now lies, and where it was once covered by water.

But the olivine tells another story as well. When it breaks down into clays, it loses a hydrogen atom that can be used by primitive microbes as an energy source. Other compounds in the drill mix were in similar states of decomposition that freed up electrons that could potentially

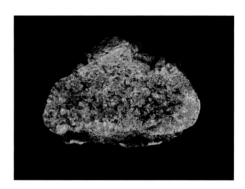

The mineral olivine is common on both Earth and Mars, especially below the surface. It is known on Earth to break down more easily than some other minerals and to easily evolve into clays.

›› On ancient Mars and perhaps even later, flowing water carved channels, created deltas emptying into water, and built alluvial fans petering out on land. Using the context camera and CRISM spectrometer on the Mars Reconnaissance Orbiter, this representation highlights the watery ancient landscape and identifies where some important minerals have been detected.

be used as microbe food. These "rock-eating" organisms, called chemolithotrophs, use electrons from inorganic compounds as their energy source.

Nothing that would conjure up pleasing images of green Martians, but rather similar to what may well have happened on early Earth. Most of the life on Earth over the eons, and even today, is and has been microbes. Our planet needed more than three billion years for them to evolve from the likes of chemolithotrophs into more complex multicellular algae and molds and, over time, oak trees and cats and humans.

But while Mars may well have started with many rather Earth-like characteristics, conditions later changed dramatically—and for the worse—in terms of a site for possible primitive life. The scientific consensus strongly holds that Mars did not have the time for complex life to evolve before it became inhospitable to it.

The other top find at John Klein was nonacidic sulfate minerals, and they, too, are most typically formed in water. The mixing of the clay and sulfate minerals was something of a surprise, since it has been theorized that Mars had an early epoch (the Noachian) when clays were formed, followed by a later period (the Hesperian) when more acidic sulfates were laid down. So finding a mixture of the two resulted in some raised eyebrows, although it is possible that both the clays and sulfates were initially formed elsewhere and later deposited at Yellowknife.

WATER YOU COULD DRINK

The icing on the cake was the detection of most of the elements associated with life on Earth: hydrogen, oxygen, sulfur, nitrogen, phosphorus, and carbon.

The findings were powerful and unprecedented:

›› **Standing water** was once present at the site.

›› **Samples** of Yellowknife's powdered rock contained clays and sulfate minerals, both formed only in water and only over long periods of time. In other words, the water had been around for a while.

›› **The clays** in particular were formed as part of a geochemical process that creates an energy source, in the form of hydrogen, that microbes could in theory live on.

›› **The water** was neither strongly acidic nor strongly alkaline, which is ideal for life.

Perhaps the most startling and eye-catching conclusion was this, as described by Grotzinger: "If this water was around and you had been on the planet, you would have been able to drink it."

Put it all together, and you had an environment that met many—though not all—of the key criteria understood to be essential to support life. That doesn't mean Martian life ever existed, but rather that based on what the scientists now understood, it sure could have. For the first time ever, scientists could declare that Mars had once been "habitable."

Before going public with the conclusion, the science team met en masse to discuss the dramatic conclusion being formulated. It was a very big deal in terms of science, but most

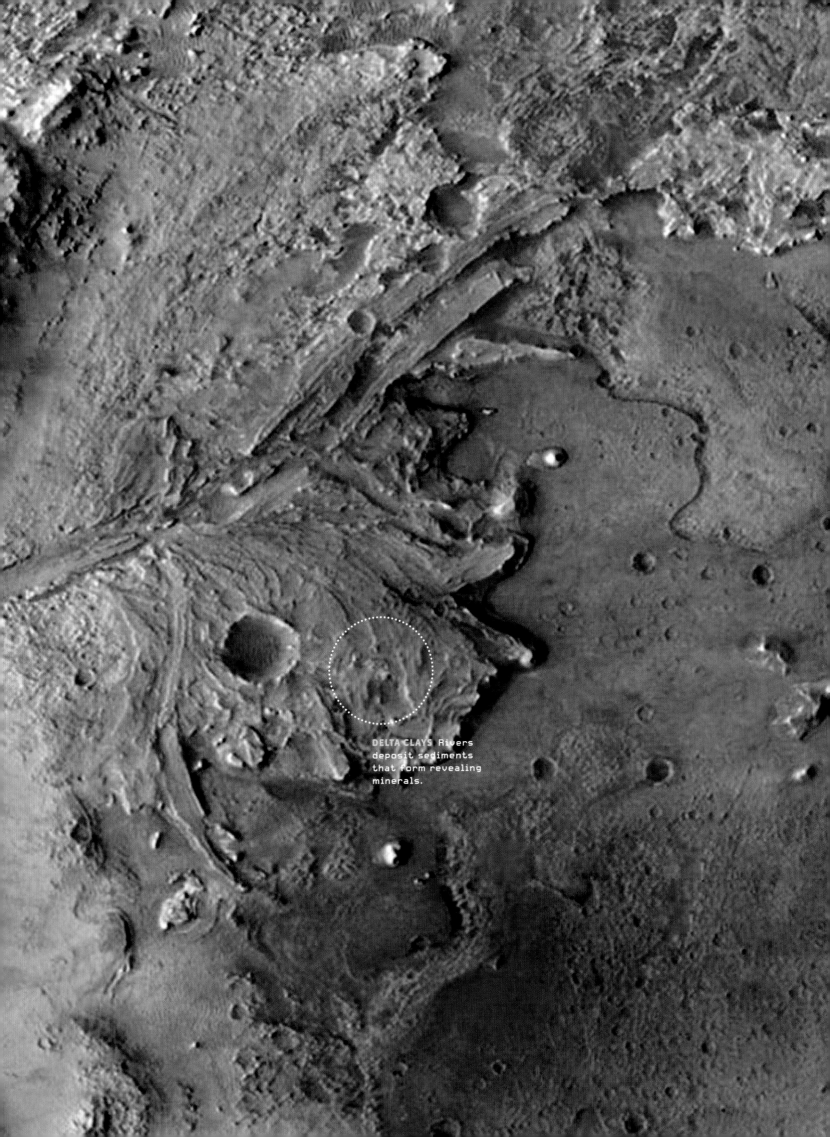

DELTA CLAYS Rivers
deposit sediments
that form revealing
minerals.

of the scientists on the extremely diverse Curiosity team readily concurred in the decision. The case was that strong.

Given that determining whether Mars once had habitable environments was one of the two main goals of the mission, a finding that those conditions did exist was, not surprisingly, a moment of triumph for the project scientist and the Curiosity team.

"Yellowknife was not our primary target, but there's no doubt that we hit the jackpot there," Grotzinger later said. "What we found is a benign environment, a place where microbes potentially could have lived and prospered."

Having declared that Yellowknife was once habitable, he also relented on using the L-word. Yes, he said, the area was probably once a pool, a low-lying playa, or a "lacustrine environment"—i.e., a lake. And, given some of the rock formations and features nearby, perhaps a rather extensive one.

The organics were still missing, but plausible arguments could be made about why they hadn't been detected. The harsh radiation suffusing the Martian surface could be breaking the bonds holding organics together, or the signature reddish and acidic iron oxide in the dust could be similarly breaking them apart. And some of the problems with the SAM contaminations and the perchlorates could have been delaying the day when they were finally identified.

Unsaid in the initial fanfare about "habitability" at Yellowknife were the implications for other areas on Mars that are known to have features that, from above, sure look like fossil alluvial fans, deltas, and possibly lakes as well.

Foreground shows familiar spare scene of Martian soils and rocks, the one encountered on previous missions. Behind is the rim of Gale Crater, one of many dramatic and science-rich features surrounding Curiosity. The Mastcam telephoto camera took the picture.

All together, you had an environment that met many
of the key criteria essential to support life.

The men and women who have identified thousands of these Martian aqueous remains have often met with resistance. And most often, those taking opposing views could point to instruments on the same Mars satellites to support their positions.

That's because in addition to their high-resolution cameras, those satellites also carried spectrometers (CRISM and OMEGA) that can read the spectral signs of specific minerals on the surface. And most commonly, those landscapes that appeared to be ancient water systems did not send back signs of the expected clays, sulfates, and other minerals formed in water.

GETTING CLOSE UP

So another of the major discoveries at Yellowknife was this: Spectral imaging from satellites is not reliable on Mars. The spectrometers that found clays and sulfates at the base of Mount Sharp did not detect any in the Yellowknife area. Apparently, the iron oxide dust that covered the flat area hid the mineral deposits from the instruments designed to measure them.

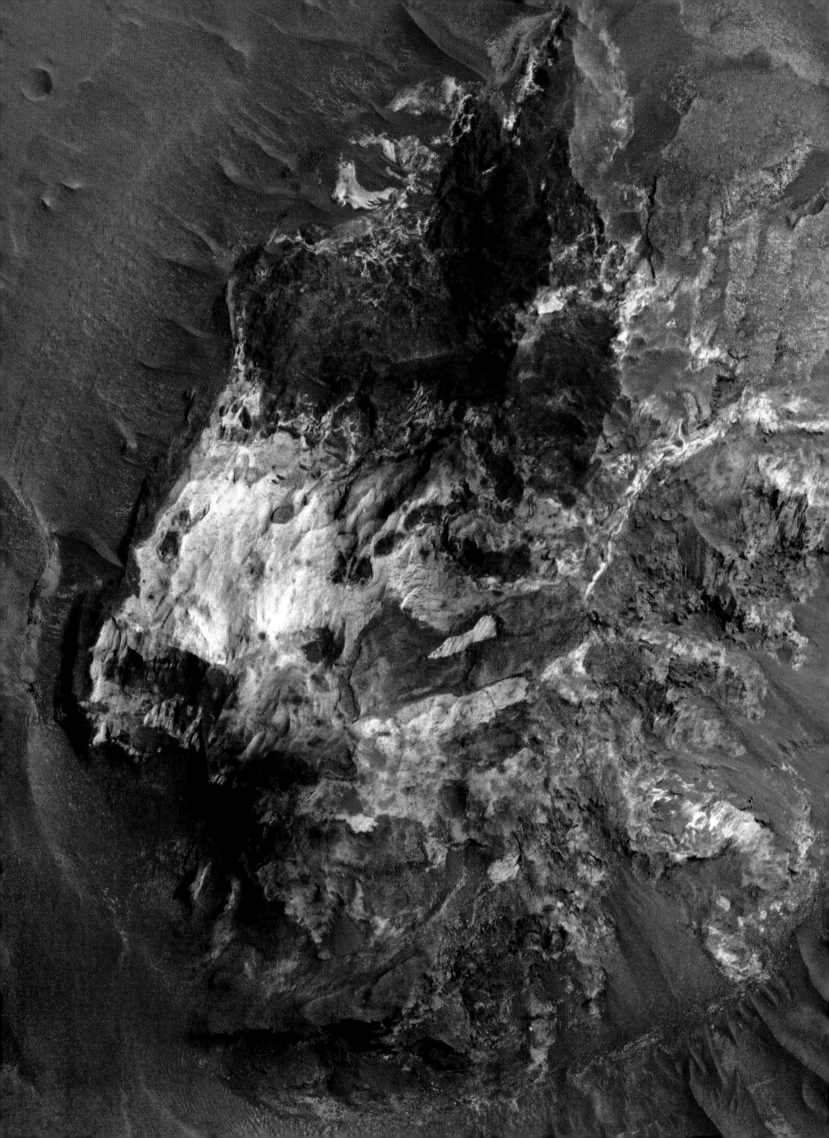

Grotzinger says the implications for Mars science are nonetheless huge.

"Nobody had any way to know there was greater than ten, fifteen percent clay down there at Yellowknife," he said. "That's a big amount of clay to have in the rocks and is way, way, way more than the detection minimum for those orbiting instruments. We know now that the dust covered the minerals in the rocks.

"This is a real shot across the bow, and I think we'll see a whole lot more about the hydrated mineral story, the clays and the sulfates. Because looking at the alluvial fans identified all around Mars, you just can't say anymore that the absence of evidence is evidence of absence. What we see instead is a spectral map of where the wind blows hard enough to blow away the dust."

And so many of those desiccated water features seen from orbit may well have hydrated minerals hidden but present. Curiosity's ground-truthing experience at Yellowknife certainly suggests that's the case.

And here's a further implication of this set of discoveries: If Yellowknife is a habitable environment based on the past presence of water, the current presence of minerals like clays and sulfates, and a source of energy associated with how the minerals are formed, then the case gets stronger for widespread conditions of habitability across Mars. If the fan and delta and lakelike formations seen from on high were in fact once active water systems, then the other conditions that contribute to habitability almost surely are sometimes present as well.

As if to underline that conclusion, the other rover still operating on Mars—the ten-year-old Opportunity—had come up with some exciting results as well. During its long odyssey on the other side of the Martian globe, Opportunity has also found clays and minerals, but they were all formed in acidic conditions inhospitable to life. Soon after the Curiosity breakthrough at Yellowknife, the older rover also discovered some clays that formed in a neutral pH environment. Two months after the Curiosity announcement, Opportunity's principal investigator, Steven Squyres, announced that the older rover had discovered a habitable Martian environment.

"This is water you could drink," Squyres said in a reprise of Grotzinger. Comparing the find to other clays identified by Opportunity, he added, "This is water that was probably much more favorable in its chemistry, in its pH, in its level of acidity, for things like prebiotic chemistry—the kind of chemistry that could lead to the origin of life."

And this second location pronounced habitable was thousands of miles away from Gale, and in an environment that had, before this discovery, been associated more with a later Mars unfriendly to neutral clays and water.

Suddenly, the case for life on early Mars had gotten rather more plausible.

Imaging scientist Justin Maki, left, and engineering manager Jordan Evans study new Hazcam and Navcam images in the Curiosity mission support center. Images from these cameras guide rover-driver decisions as well as give early looks at the passing landscapes.

<< Many of the large outflow channels from craters and other features on Mars have been attributed to what are termed catastrophic releases of groundwater, since they seem to begin in areas where the ground appears to have collapsed.

MARS PRESENT

« DRY AND COLD

The alluvial fan etched
into bedrock, raised ridges
within the fan, and outcrops
of sedimentary rock with
mineral contents hint at
a once different climate
at Peace Vallis.

THE ANOMALY

CHAPTER 8

The night a software glitch almost took Curiosity down

MAGDY BAREH is an electrical and computer engineer who specializes in things that can go wrong with the
Curiosity main computer software. He earned his expertise
the hard way: He helped build the guts of the rover's twin
main computers, he designed much of the software that
controls how it behaves within the system, and he also
was part of the team that built the rover's fault-protection
system. That most essential and complex function consists
of digital trip wires that autonomously stop computer
glitches from cascading into big problems.

After seven years of focus on the core of the Curiosity computers, he had the inner workings of the machines basically imprinted into his brain.

By February 27, or Sol 200 of the mission, Bareh had settled
into a routine of monitoring the data (or telemetry) coming
down four times a day from the rover's computer system.
The data came directly to his JPL e-mail account and generally presented nothing to worry about.

A half year into the mission, a commonly heard refrain
was that the biggest "anomaly" for Curiosity—the most
unusual characteristic the rover presented—was the near
total absence of anomalies. Virtually all systems were
working as intended, if sometimes in need of tweaking.
A sieve involved with delivering the proper-size grains of
Martian soil and rock powder, for instance, had been found
in an Earth test bed to be potentially weak. The result was

that the sieve on Mars had to be used more gently—no "thwacking" to clean it off—but it was hardly a showstopper.

Bareh was in a meeting when the 10 a.m. data came down in the e-mail with a warning that looked a little strange. It was an error involving several bits, the smallest measure of computer information. Bit errors happened all the time and were no big deal.

A while later he strolled over to the mission support area and asked some of the engineers on the consoles what was going on. For as yet unknown reasons, the reported error seemed to be associated with a failure of the rover to send higher-level data back to JPL properly, although the low-level telemetry was arriving. Puzzling . . . Bareh tried to think it through.

What does that absence of files have to do with a memory error in these warnings? There was every indication that all the data should have arrived, but very little had arrived . . .

Then Bareh and a still small group of others learned that something far more troubling had occurred: A second orbiter pass sent back the information that while the computer was supposed to have shut down almost all rover activities by noon (late afternoon Mars time), it hadn't. Some instruments were deployed—hanging out in the Martian breeze— and some systems that needed rest were running instead. One of the core functions of the computer just didn't seem to be working.

Why didn't it shut down? Bareh asked himself more than once. Yet the inability to shut down wasn't the big problem; it was not knowing why it was happening that was worrisome.

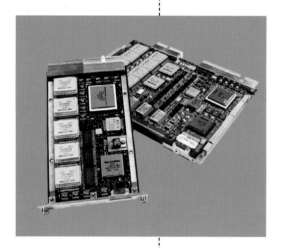

The RAD750 central processor unit was used in both Curiosity main computers. All parts of the computers are hardened to withstand radiation, but the protection isn't foolproof.

How can we trust this computer?

Well, we can't.

Bareh was now officially concerned, as were a growing number of others around him. Curiosity officially had an "anomaly," and it seemed to be a potentially big one. Bareh's head filled with scenarios of what might be going wrong, and each was more ominous than the one that came before it. The endpoint was the grim fear—and potential reality—that the rover's computers would not be able to respond to commands from Earth . . . perhaps forever.

SWAT TEAM ASSEMBLES

Calls and messages started going out to Curiosity computer hardware and software experts, and an "anomaly resolution team"—a kind of bomb squad pulled together when a serious problem arises—was quickly convened, with Bareh as leader.

Over the next 24 hours a little-known drama played out, with the fate of the mission very much in the balance. And the fast-moving anomaly brought home one of the constant tensions of this mission, and of missions similar to it. It's usually described as the sniper scenario.

The failure to shut down highlighted the always tenuous nature of Mars exploration, but it made clear that the vast and complex autonomous fault-protection system on Curiosity is, despite the problem, robust. What's more, it put onto center stage people like Bareh and a team of Curiosity

"The mission was at risk and we were extremely concerned, but we were confident in the team's ability to recover Curiosity."

Because Bareh and his systems-engineering colleagues helped design and develop the computers that run Curiosity, he knows the complexities of the equipment inside and out. And as leader of the anomaly team that's dispatched when something goes wrong with a computer, he is always plugged in and on call. Whatever the challenge, and however he may feel on the inside, Bareh projects cool and calm. He says that acting otherwise is harmful to resolving whatever has gone wrong, and makes essential teamwork more difficult. Trained as an electrical and computer engineer, he has also worked on the Spitzer Space Telescope and the Dawn mission to an asteroid and comets.

Magdy Bareh, Jet Propulsion Laboratory, NASA

A fault-protection engineer who helped develop Curiosity's two main computers, Bareh stepped in as anomaly team leader when the rover's main computer failed and led the emergency response that saved the vehicle (artist's rendering below).

veterans including fault-protection specialist Tracy Neilson, and software engineers Rajeev Joshi and Daniel Gaines (who performed a speed autopsy of the corrupted data-management system).

From their standing start that Wednesday morning, they quickly diagnosed enough of what was wrong 150 million miles away to know the computer was no longer dependable. They made the risky but essential switch to the backup main computer as a most worrisome deadline approached, then waited breathlessly to learn whether the change had succeeded. And they did it all in less than 24 hours.

What happened?

Normally, Curiosity is programmed to wake around 9 a.m. Mars time and operate until midafternoon. Then the equipment is instructed to automatically shut down until the middle of the night, when it is awakened again. The orbiting Mars satellites are in range at that time, and much of the rover's data—photos, science data, and information about the well-being of the vehicle—can best be transmitted then.

The afternoon's commanded sleep is especially important in order to recharge the battery and, as a result, keep the instruments warm during the frigid night. Fault-protection specialists designing the computer core identified a slew of possible sleep problems for the rover— "Insomnia," "Sleep Apnea," "Too Busy to Sleep," "Narcolepsy"—and had built in autonomous software triggers to shut down the rover in case they appeared.

Initially, the engineers and mission managers suspected cosmic rays that hit a particularly sensitive part of the main computer directory. These highly charged particles pierce the rover shell all the time without significant damage, but perhaps this encounter occurred in an unusual way.

On Sol 200 the Curiosity team declared a space emergency based on a small but hazardous software and hardware "anomaly," as they called it. The rover was still in the Yellowknife Bay area (above) when the problem became known, setting off a tense drama to repair it before it caused lasting damage.

Bareh was now officially concerned. Curiosity had an "anomaly," and it seemed to be a potentially big one.

JPL's Gerald Clark inspects Curiosity's twin in the Vehicle System Test Bed, or VSTB, the proofing ground for evaluations of the mission's hardware and software.

Whatever the root cause, the result was that the main—or A—computer got hung up in the way computers do, but would not allow for the shutdown that would lead to a rebooting. And while computer users on Earth are quite accustomed to rebooting or unplugging as a way to get their machines working again, switching to a backup computer on Mars was a very big deal: It usually involves a flight director going to the JPL or NASA headquarters program office for institutional approval. But it quickly became apparent there was no time for that.

TURNING TO THE TEST BED FOR ANSWERS

One of the first people brought into the anomaly team was fault-protection specialist Tracy Neilson, who knew backward and forward all the autonomous actions that the rover's computer could and would take to keep itself safe. Like Bareh, she had been on the rover team for many years and helped design the system.

She also had experience with sudden adversity as fault-protection leader for Spirit and Opportunity, the two Mars exploration rovers that landed on the planet in 2004. She had played an essential role in figuring out what had caused a somewhat similar, though much more lengthy, near-death experience with Spirit that began 18 sols after landing on the planet.

Neilson had actually left the Curiosity team six days before the anomaly and was looking forward to a going-away party that evening. But she was still one of the relatively small number of people who got e-mails four times a day with "event reports" from Curiosity, rundowns on the state of its central systems.

So she had also seen the initial problem e-mail from Mars, and got a few text messages from colleagues asking for her thoughts. But she didn't see any immediate problem and continued with the presentation she was giving related to her new position and tasks.

That approach was entirely in keeping with the "If it's not broken, don't fix it" philosophy of the mission to that point. Curiosity is a very complex creation, and Neilson knew that trying to fix something from millions of miles away could cause more headaches than improvements.

But the deciding text that came in was from Bareh, her colleague on Curiosity. His message to her was delivered with the usual JPL calm, but it was nonetheless alarming, to say the least. His message: The situation might not be at the Sol 18 crisis level, but it was serious and not well understood at all.

Neilson immediately left her meeting, her scheduled presentation, and her going-away party, and headed back to Curiosity mission support.

During her first anomaly meeting, she learned the outlines of what had gone wrong and the small group's understanding of why. Then Neilson did what planetary engineers always do to try to better understand a problem: She went to the test bed and tried to replicate the problematic behavior.

MODELING FAILURE

JPL has Curiosity's twin in its In Situ Instrument Lab—a large room with a sandy, Mars-like floor, retired test versions of past rovers, and a bank of computer racks fed by a Medusa's head of wires and cords wrapped into thick black casings.

The test-bed computers are nearly identical (in function, though certainly not appearance) to the ones on Curiosity, and nearby monitors can communicate with the replica computers. By introducing similar faults into the test computers, they could diagnose the errors they were seeing on Mars.

Curiosity is a very complex creation, and trying to fix something from millions of miles away could cause more headaches than improvements.

ALL SYSTEMS NOT GO Bobak Ferdowsi and Bradley Compton monitor a screen in the mission support center. The rover's computer was still able to communi-

Within several hours, Neilson, Jonathan Grinblat, and Gene Lee were able to reproduce the fault causing so much trouble. Essentially, the computer could not go into safe mode or shut down, either when commanded or on its own. In Curiosity engineering terms, the "shutdown had gotten locked out."

Why? Because the computer continued to "stroke the watchdog," which meant that the mechanisms that autonomously turn off the computer were getting the signal that all was well. The rover has an electronic system designed to awaken the backup computer if the main computer fails, but the main computer's software didn't diagnose that its own system was misbehaving.

To test the strength of the remaining fault-protection system, Neilson, Lee, and Grinblat injected a fault to trip the autonomous fault protection. The result was encouraging in that the underlying fault-protection system remained in operation. But it was also worrisome, because the fault-protection system could inadvertently turn off the radio in this hung-up state. That meant the rover was just one additional fault away from losing communication with Earth, possibly forever.

Back at the mission support center, an endless series of meetings was beginning to make sense of how the fault had occurred, or at least where in the system it was located. (It turned out to be in a portion of the catalog, or directory, of the computer's bulk data storage memory, analogous to a hard drive in a personal computer.) In a now overflow meeting in the fishbowl meeting room just off the support area, top JPL and Curiosity engineers joined the anomaly team, the mission managers, and others with expertise that might be helpful.

By late afternoon, the anomaly team was unanimous in concluding that the main computer was unstable and getting more so. It was too risky to try a reboot because they just didn't know what it would do. And so the only real option was to switch to the backup.

The Space Flight Operations Center at JPL was the early hub for all Curiosity activity on Mars. As the mission progressed, the main action moved to smaller areas within the sprawling campus in La Cañada Flintridge, outside of Pasadena.

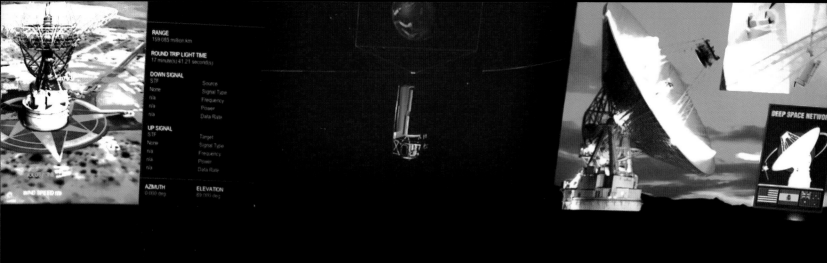

Curiosity would autonomously change configurations. In its confused state, the system could lock out all future commands coming from Earth.

Sol 201
09:17:48 LMST
2013-059 12:00:11 UTC
Feb 28 04:00:11 PST

WAITING FOR WORD In the early hours of February 28, Curiosity computer specialists commanded the rover to turn off its main A computer and switch to the backup B computer. They would wait for hours before learning whether the switch succeeded.

That path also carried risks—a kind of commanding in the blind—but it appeared to offer better odds of success. There was no time for the usual lengthy and considered review. Not only was the main computer unstable, but later that night the telecommunications system on Curiosity would autonomously change configurations, a deadline that the team on Earth had no control over. In its confused state, the system could lock out all future commands coming from Earth.

UP ALL NIGHT

It was night on Mars, and some 80 degrees below 0°C. At a 5 p.m. meeting in the fishbowl, mission managers wanted to know whether the rover's rechargeable lithium-ion battery had enough juice to keep the science instruments warm at night. The battery remained sufficiently charged, though it would certainly lose power over a few days.

As for communication, the limited but tried-and-true X-band radio antenna and transponder were down for the night. Two Mars satellites would pass over Curiosity before dawn and could theoretically send far more detailed data, but sending complex and not fully tested commands seemed especially risky now.

During the meeting, Lee and Grinblat continued to work in the test bed to come up with commands for turning off the main computer and activating the backup. This required sending two commands that put the main A computer into an extreme isolation mode, where it would sit and essentially do nothing. This was different from sleep mode, where the computer remains connected and interacting with other Curiosity systems. Their goal, as Nielson put it, was to tell the computer to "go spin."

If the short set of commands was received and understood by the hardware, then the fault-protection system would also turn on the backup computer, which would then automatically step in to take over control of the rover. Assuming, of course, the backup was working properly.

Ironically, the backup B computer had been scheduled for a complete software upgrade that week. It hadn't been used since before the spacecraft had landed, and that meant many of the software upgrades based on actual Curiosity experience with the terrain, the drill, and the cameras had never been uploaded. The upgrade had been canceled because the science coming in had been so exciting and the main computer was working so well.

As the anomaly team plan became more solid, it was shared with more and more of the Curiosity (and ultimately JPL and NASA) managers and decision makers. As Bareh put it, "Fear had gone to the highest levels, and the deadline was looming." That's why, when a climatic midnight meeting began in the fishbowl, the several dozen people present (with many more listening in by phone) were all pretty much on the same page. The best odds for getting out of their mess was through the inherently risky switchover, and the issue was not whether to implement the move, but how.

"I usually do a lot of explaining in meetings like that, letting people know what we're doing and exactly why," Neilson said. "But for this meeting, I was pretty silent because everything was clear. If someone had raised an objection, I would have jumped in for sure. But that didn't really happen."

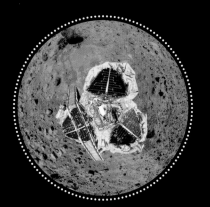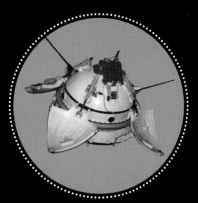

Curiosity Rover
@MarsCuriosity

Following

Don't flip out: I just flipped over to my B-side computer while the team looks into an A-side memory issue go.nasa.gov/ZN8xsx

↩ Reply ⇄ Retweet ★ Favorite ••• More

413
RETWEETS

125
FAVORITES

12:25 AM · Mar 1, 2013

NOT ALL SYSTEMS GO ›› Most missions to Mars fail, and even those that succeed do so by overcoming inevitable problems that arise. Of the 45 missions sent to Mars from Earth, about one-third completed their tasks. American rovers have been stalwarts, with both Sojourner (above, left) and Spirit (right) operating far longer than their prescribed missions. The Mars 3 lander (center), launched in 1971 by the Soviet Union, was the first to achieve a soft touchdown on Mars. Communication with Earth, however, was lost after 14 seconds and never regained. Curiosity, by contrast, is highly communicative—and social media-savvy (left).

As those minutes ticked by, Neilson "felt the weight of the world" on her shoulders.

STANDING BY Deputy mission manager Jennifer Trosper and Keith Naviaux wait anxiously during the wee hours of February 28 for word about Curiosity's fate. The mission was facing the possibility of a sudden and completely unexpected end.

OVER TO COMPUTER B

It was about 30 minutes after midnight on Earth when the switchover decision was made and a formal space emergency was declared. The plan was to send the switchover signal via X-band three or four hours later.

The commands would go to Curiosity via the Deep Space Network, three interconnected radio telescope dishes and antennas in Canberra, Australia, outside Madrid, Spain, and just north of the Mojave Desert town of Barstow, California. Together they make possible 24-hour, interplanetary telecommunications directly to and from Earth, a capacity that is not available through any other system.

A portion of the electromagnetic spectrum called the X-band has been set aside exclusively for Deep Space Network use, and most NASA spacecraft are launched with antennas and receiver/transmitters that can talk over the network. Battle-tested and reliable, Curiosity's X-band antenna sent back information even during the most intense phases of the descent and landing.

The way the team would know that it had successfully put the main computer into isolation and activated the backup would be through what is called a 51-minute "tone" or "whistle" coming from the X-band. That so-called "sound" is the signature of a computer in safe mode, but it actually produces no noise at all. Rather, it produces a visual frequency signal of a particular strength that comes in as part of a data dump and basically says, "I'm here." That's what everyone would be watching for.

But first the command had to be sent through the network, and on a day of mishaps another problem arose. The station best situated to relay the command was Madrid, and after the declaration of a spacecraft emergency the technicians there were obliged to drop whatever else they were doing and attend to the problem. But Madrid was experiencing a snowstorm that night, and the antenna might be going down at any moment. Another headache and more tension as the deadline approached.

That was about 4 a.m. Pasadena time and 8 a.m. on Mars. If the command succeeded, then at 9 a.m. Mars time the "tone" would turn on. So the worried, hopeful, exhausted team would find out within the hour if the first part of their plan had been successful: getting the main computer to finally shut down in isolation mode. They wouldn't know for sure the answer to the equally essential question of whether the backup computer had turned on properly until about 10 a.m., when additional data would come back via one of the orbiting satellites.

About 50 people gathered in the room and waited. A 19-minute Mars-to-Earth communications delay had been factored in; more waiting. Then the time came for the signal to arrive, and there was nothing. As minutes passed and tension grew, a hush settled on the team and the room fell utterly silent.

MOMENTS OF SUSPENSEFUL WAITING

Losing a spacecraft or rover can be, and usually is, devastating to the team responsible for it. The loss includes a vanished

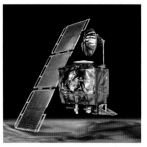

TRIED AND FAILED The Mars Climate Orbiter (above, left and right) arrived at Mars in 1999 but did not properly enter orbit and disintegrated.

opportunity to learn about a planet or comet or solar system, inevitable criticism and second-guessing of NASA, and often a reshuffling of roles at JPL and the agency. NASA lost two
Mars missions in the late 1990s—the Mars Polar Lander and the Mars Climate Orbiter—and
a team was brought in to revamp the entire Mars program as a result. Space exploration is
inherently very risky. Yet a mind-set nonetheless remains that "failure is not an option"—an
adage made famous in the movie that chronicled the rescue of the crippled Apollo 13 spacecraft and its astronauts.

For Neilson, a loss went one step further—to personally letting down not only NASA and the
mission team, but the entire nation. Curiosity had been launched and landed on Mars with such
acclaim, and to get terminally hung up before attempting some of its most daring exploits would
be crushing to the millions following the rover's exploits. As those minutes ticked by, Neilson "felt
the weight of the world" on her shoulders.

Just as Neilson and her expectant teammates were beginning to harden their emotions
for a blow that seemed headed their way, a blip showed up on the monitor.

Then the time came for the signal to arrive, and there was nothing.
A hush settled on the team and the room fell utterly silent.

A WAITING GAME Magdy Bareh, second from
left, leader of the anomaly resolution team
that diagnosed the problem from afar and
created a possible solution, waits to see
if the fix worked. Looking worried, right, is

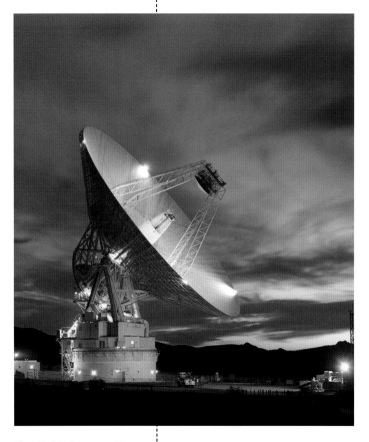

The Goldstone radio
telescope in the
Mojave Desert is the
American contribu-
tion to the Deep
Space Network. With
radio telescope
transmitters and
receivers around the
world, the network
always has at least
one available to
operating spacecraft.

Was it the "tone"? It was tiny on the screen, and initially hard to make out. But it was the right frequency at the right strength.

The command had worked; the communicating computer was in safe mode and was able to say so to Mission Control.

Relief swept the room, with whoops and hollers, hugs and more than a few tears. It was a celebration not unlike the joyous scene at landing, except that only a relative handful of people at JPL understood the disaster that had apparently just been averted.

Telecommunications team leader Peter Ilott spoke for many of the drained but now ecstatic revelers: "I really thought the MSL luck had run out."

The next morning, the ever composed mission manager Michael Watkins summed up the previous day. The anomaly, he said, "Came very close to killing the mission . . . There were things going on that, if not dealt with quickly, would have been fatal, or certainly could have been fatal."

FIGURING OUT WHAT HAPPENED

Much of the team headed home in the still dark early morning for a few hours' sleep, relieved that the crisis appeared to be over but still worried about the health of computer B (now, they hoped, the main computer) and what had mysteriously corrupted computer A.

So Neilson and many others jammed back into the fishbowl not many hours later for a 10:30 a.m. report on the state of the computers as reported in a 10 a.m. pass over the rover by the Mars orbiter Odyssey.

The news was good: It was indeed computer B (the backup) that had reported in earlier that it was now in safe mode. That meant computer A (the former main computer) was finally shut down and in isolation. The crisis did appear to be over, though several complicated weeks of analyzing, patching, and updating followed before the rover did much of anything again.

During that time, the software team wrote and sent to Mars commands that would patch the troubled section of the A computer and better protect the B. To make sure neither computer would ever get caught up in a shutdown loop again, they also patched the code to ensure that if the computer ever stays awake for 30 hours, then automatically—whatever other messages it was getting—it would reset into isolation mode.

Meanwhile, a larger "tiger team" of computer hardware experts got to work analyzing what had caused the near-fatal problem. The team brought in hardware and software specialists and put together what its leader, Curiosity chief avionics engineer Jim Donaldson, called a "fishbone."

A fishbone is a convenient way to diagram the structure of a computer system in terms of the branches connecting to the main computer backbone, and the team initially came up with 38 possible causes of the anomaly to put onto the skeleton. Each is added to the diagram and is called a "leaf," a very mixed metaphor that nonetheless makes sense to the computer experts. Then, systematically, they sought to eliminate each one of those causes.

Donaldson, who had been in Hong Kong when the anomaly became known and participated in the meetings by phone until his return the next day, says that the architecture of the computers allowed his team to probe and analyze the corrupted computer using hardware alone.

With the A computer in debug mode, Donaldson's tiger team was able to, in effect, "put the computer on an operating table" and have at it. This was done from Earth, with the B computer serving as the doctor who gathered information from the diseased hardware "organs." That data was then downlinked to engineers eager to start the diagnosis.

That meant the rover was one additional fault away from losing communication with Earth, possibly forever.

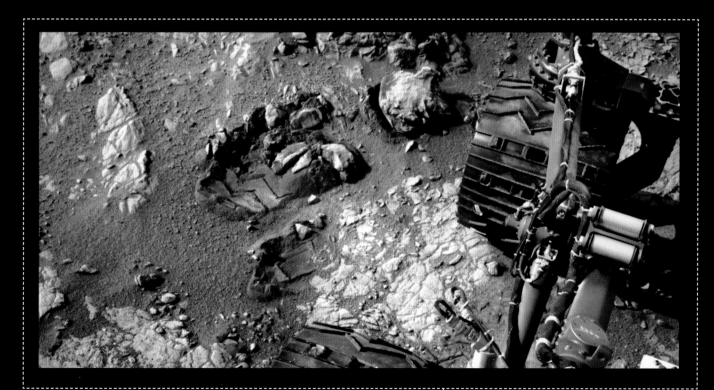

REBOOT ON B After the switch to the B computer, all the systems it now controlled needed a rigorous health check, including the essential Navcam engineering camera, an image from which is seen here.

Among the first things they learned was that while the computer's software did get corrupted, this fault was not the root cause of the problem. Even with the software not running, the same problems surfaced in the operating-room patient, and so software was taken off the fishbone.

It took longer, but over the weeks cosmic rays were also largely removed as a root cause. A fault set off by a cosmic ray has a particular residual signature—the system behaves a certain way afterward when asked to erase, write, and read information—and the patient on the operating table responded differently. Also, the rover's radiation-detection instrument, and other radiation sensors in space, had not detected any unusual activity at the time of the problem.

ROOT CAUSE

Additional testing revealed that the problem only affected an individual hardware chip in computer A. With the identification of a single failed part, the focus of Donaldson and the tiger team moved from software to hardware to understanding how the part, which was not close to its specified life expectancy, could have died.

It could have been a single soldering job, or a single relay between two points, or a single malfunctioning pin that resulted in the failure of a single chip. Donaldson says all the computer boards on Curiosity had been tested and retested before launch, and nothing akin to the anomaly fault showed up.

As they sought to understand the culprit hardware failure in the JPL test bed, they vibrated, thwacked, and froze the computer boards to simulate what was happening on Mars as the drill and the sample collector were put to use while the computer simultaneously dealt with daily temperature swings of 100 degrees Fahrenheit and more. Again, no problem.

"So basically," Donaldson said, "we have a situation where it sure looks like a single small piece of hardware failed well before its specs would say it's in danger, and failed for reasons we don't fully understand." The phenomenon, he says, is called "infant mortality."

In the ensuing weeks and months, Curiosity engineers were able to pinpoint where that early mortality had occurred: They isolated the memory area. They sent up commands that allowed the A computer to run differently, and ultimately they certified that it was in good running order and could take over again if needed.

Meanwhile, it became clear that while software was not the root cause of the anomaly, the hardware failure had highlighted a serious flaw in the software. As Bareh explained, "The software should have been able to handle the hardware problem, should not have gotten hung up like it did. So we had a hardware problem that exposed a software problem, and the two together put the mission at serious risk."

The metaphorical sniper had taken a clean shot and almost took out the rover. But thanks to quick and effective thinking from 150 million miles away, the patient was successfully stabilized, the wound was located and treated, and the team announced a full recovery.

The sniper, however, remained on the loose. He comes with the territory.

MODELING THE PROBLEM >> System engineers used this "fishbone" model to track
their ongoing analysis of the anomaly. Its parts show the testing and thought
processes that went into determining what had gone wrong with the A computer,
leading to the answer: A small hardware failure led to a larger software failure.

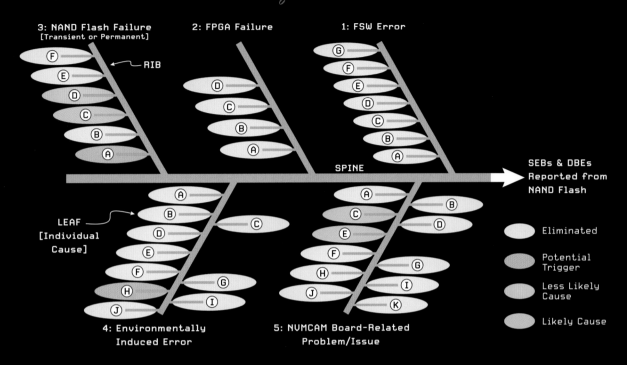

3: NAND Flash Failure
[Transient or Permanent]

2: FPGA Failure

1: FSW Error

RIB

LEAF
[Individual
Cause]

SPINE

SEBs & DBEs
Reported from
NAND Flash

Eliminated

Potential
Trigger

Less Likely
Cause

Likely Cause

4: Environmentally
Induced Error

5: NVMCAM Board-Related
Problem/Issue

"So we had a hardware problem that exposed a software problem,

and the two together put the mission at serious risk."

ALL SYSTEMS ONCE MORE GO There was
jubilation in the mission support center when a small blip appeared on the
monitor telling the relieved team at
JPL—including Rick Welch, center—that
the computer switch had succeeded.

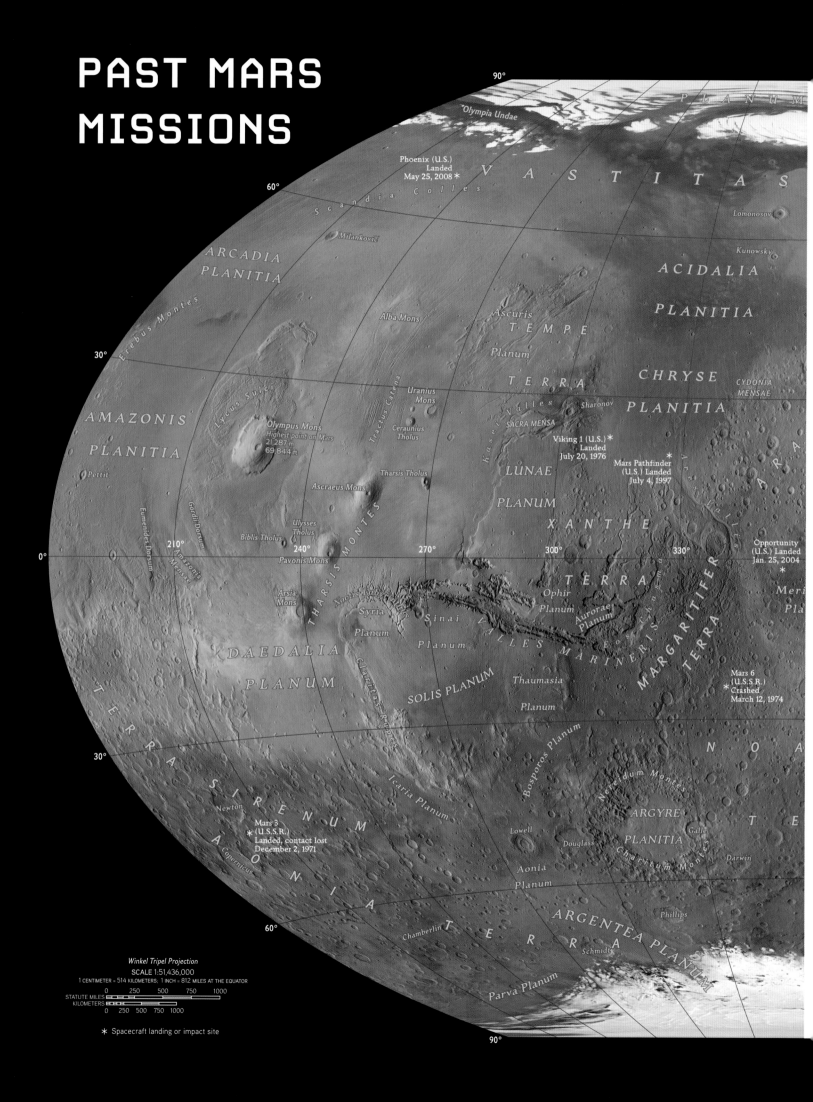

PAST MARS MISSIONS

Spacecraft landing or impact site

Winkel Tripel Projection
SCALE 1:51,436,000
1 CENTIMETER = 514 KILOMETERS; 1 INCH = 812 MILES AT THE EQUATOR

STATUTE MILES 0 250 500 750 1000
KILOMETERS 0 250 500 750 1000

Phoenix (U.S.)
Landed
May 25, 2008 ✳

Viking 1 (U.S.) ✳
Landed
July 20, 1976

Mars Pathfinder ✳
(U.S.) Landed
July 4, 1997

Opportunity
(U.S.) Landed
Jan. 25, 2004 ✳

Mars 6
(U.S.S.R.)
✳ Crashed
March 12, 1974

Mars 3
✳ (U.S.S.R.)
Landed, contact lost
December 2, 1971

Olympus Mons
Highest point on Mars
21,287 m
69,844 ft

ARCADIA
PLANITIA

AMAZONIS
PLANITIA

VASTITAS

ACIDALIA
PLANITIA

TEMPE
TERRA
Planum

CHRYSE
PLANITIA

CYDONIA
MENSAE

LUNAE
PLANUM

XANTHE
TERRA

TERRA

Ophir
Planum

MARGARITIFER
TERRA

VALLES MARINERIS

DAEDALIA
PLANUM

SOLIS PLANUM

Thaumasia
Planum

ARGYRE
PLANITIA

SIRENUM

AONIA

Aonia
Planum

ARGENTEA PLANUM

TERRA

TERRA

Olympia Undae

Scandia Colles

Milankovič

Lomonosov

Kunowsky

Alba Mons

Ascuris

Sharonov

Kasei Valles

SACRA MENSA

Uranius
Mons

Ceraunius
Tholus

Tharsis Tholus

Ascraeus Mons

Ulysses
Tholus

Biblis Tholus

Pavonis Mons

Arsia
Mons

Noctis Labyrinthus

Syria
Planum

Sinai
Planum

Aurorae
Planum

Pettit

Lycus Sulci

Erebus Montes

Eumenides Dorsum

Gordii Dorsum

Amazonis
Mensae

Tractus Catena

THARSIS MONTES

Claritas Rupes

Icaria Planum

Bosporos Planum

Nereidum Montes

Lowell

Douglass

Galle

Darwin

Charitum Montes

Newton

Copernicus

Chamberlin

Schmidt

Phillips

Parva Planum

210° 240° 270° 300° 330°

0°

30° 30°

60° 60°

90° 90°

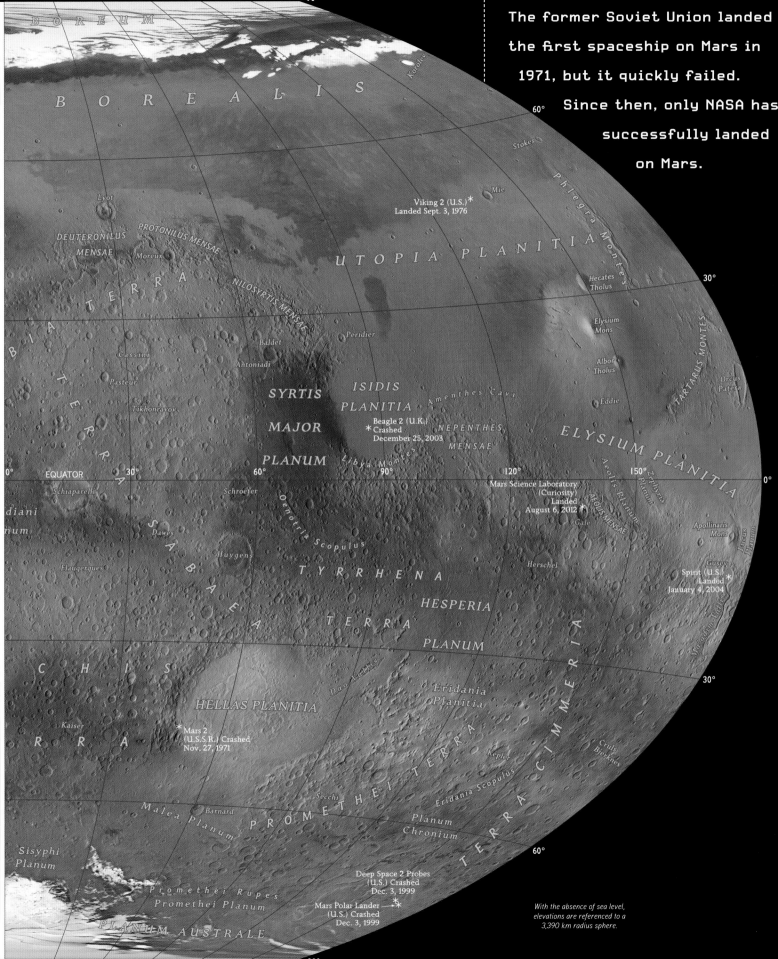

The former Soviet Union landed the first spaceship on Mars in 1971, but it quickly failed. Since then, only NASA has successfully landed on Mars.

BOREUM

BOREALIS

60°

Korolev

Stokes

Phlegra Montes

Lyot

Mie

Viking 2 (U.S.) *
Landed Sept. 3, 1976

30°

DEUTERONILUS
MENSAE

PROTONILUS MENSAE

Moreux

UTOPIA PLANITIA

Hecates
Tholus

BIA TERRA

NILOSYRTIS MENSAE

Peridier

Elysium
Mons

Cassini

Baldet

Albor
Tholus

Orcus
Patera

Pasteur

Antoniadi

ISIDIS
PLANITIA

Amenthes Cavi

Eddie

TARTARUS MONTES

Tikhonravov

SYRTIS

MAJOR

Beagle 2 (U.K.)
* Crashed
December 25, 2003

NEPENTHES

MENSAE

ELYSIUM PLANITIA

0°

EQUATOR

30°

60°

90°

PLANUM

Libya Montes

120°

Aeolis Planum

150° Zephyria
Planum

0°

Schiaparelli

Schroeter

Oenotria Scopulus

Mars Science Laboratory
(Curiosity)
Landed
August 6, 2012 *

Aeolis Mensae

Gale

Apollinaris
Mons

diani
num

Dawes

Huygens

Herschel

Gusev

Flaugergues

TYRRHENA

Spirit (U.S.) *
Landed
January 4, 2004

SABAEA

HESPERIA

TERRA

PLANUM

CHIS

HELLAS PLANITIA

Dao
Eridania
Planitia

30°

Kaiser

* Mars 2
(U.S.S.R.) Crashed
Nov. 27, 1971

Kepler

Cruls
Bjerknes

R   A

Secchi

Eridania Scopulus

Malea Planum

Barnard

PROMETHEI TERRA

Planum
Chronium

Sisyphi
Planum

Deep Space 2 Probes
(U.S.) Crashed
Dec. 3, 1999

Promethei Rupes
Promethei Planum

Mars Polar Lander
(U.S.) Crashed
Dec. 3, 1999

PLANUM AUSTRALE

60°

*With the absence of sea level,
elevations are referenced to a
3,390 km radius sphere.*

STORIES
IN STONE

<< The rover passed these rocks, dark and shiny sentinels, on the way from Yellowknife Bay to Mount Sharp. Most likely consisting of basalt, they could have been delivered to Gale Crater via volcanic explosion or a meteorite impact. Gravity on Mars is only 38 percent of what it is on Earth, which means heavy objects can travel much farther.

CHAPTER 9

The more we know,
the more we wonder
about what the rocks
are saying

THE ORIGINAL PLAN for exploring Gale Crater was for Curiosity to leave Bradbury Landing about Sol 20 and head to Mount Sharp. But it was Sol 324 when the rover finished up around Yellowknife Bay and began that long trek to the mountain.

The path to Mount Sharp was dubbed the Rapid Transit Route. A few months before, during a raucous general science team meeting at Caltech, the group had voted to drive as expeditiously as possible to what was, after all, the primary destination of the mission. But they were hardly unanimous.

The team split almost evenly into three groups: the Mount Sharp or Bust group; those who wanted to spend considerable time along the way to Sharp examining outcrops and dunes and the chemical makeup of compelling samples they might find; and those who wanted to camp out longer still at Yellowknife. After all, Yellowknife certainly was providing a steady stream of important discoveries, with hints of much more to come.

But the science team leaders who design the strategy of the mission—the long-term planners and project scientists—concluded in the weeks that followed the science-team vote to go for a slightly modified Mount Sharp or Bust. Most of the time the rover would be driving, stopping only for rest and when staffing on Earth required restricted use of the rover. Five stops were identified where ChemCam,

APXS, and maybe even CheMin and SAM would return to action if the rocks and landscape looked sufficiently important or mysterious.

With the nine-kilometer (six-mile trek) to Sharp before them, it was time for the science team to more fully analyze and report on what had already been revealed about Mars. These findings involved the Martian atmosphere, winds, and climate, but most broadly they focused on what the rocks and soils of Gale Crater were telling them.

What had been discovered about the chemical makeup of the rocks? What had been learned about the placement of rock units and formations in the landscape? What was the layering of the rocks telling them? And perhaps, most exciting, what had the contents of the scoop and drill samples revealed?

Beyond these findings was the challenge of putting the new information together and coming up with a credible story about what had happened over the eons at Gale Crater. And especially, what would that story have to say about the possibility of ancient life on Mars?

What the rocks had revealed so far included that:

Black basalt in the Minute Granite Gorge of the Grand Canyon is from lava flows, similar to those on Mars.

>> **The ever present** Martian dust and soil has significant amounts of water in it—1.5 to 3 percent. Martian scientists knew that water, as identified by the presence of hydrogen, existed in some Martian soil. But the ChemCam laser zapper nailed the water-dust connection by finding a strong signature of hydrogen on every surface it tested, and then SAM measured the amounts.

>> **A kind of rock** that is rare but well studied on Earth turned up in the early days of Curiosity's exploration. Named after Jacob "Jake" Matijevic, the JPL chief surface systems engineer who died two weeks after Curiosity landed, the dark-colored, triangular rock was identified as a mugearite, a type of rock typically found on ocean islands on Earth. Jake M was made from alkaline, possibly water-rich, molten rock deep underground—definitely not conditions generally associated with Mars.

>> **Water was present** when the mudstones, sandstones, conglomerates, and clay and sulfate mineral deposits were laid down in Gale Crater. Without water, they could not have been formed. And the water was not there in only one period, but likely in many. This could be seen in the veins of calcium sulfate deposited in rock fractures and the mineral bumps (concretions) on other surfaces.

>> **The rock story** of Gale Crater was largely a sedimentary one. Molten material that shot up from below the surface and formed the original bedrock was broken down and modified by wind, water, extreme temperatures, crater impacts, glaciers, and more. These released sediments would move easily around, especially into the bowl of a crater, and

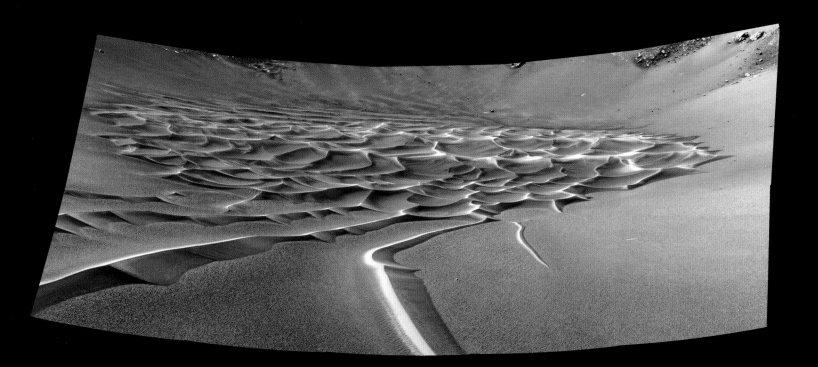

MARTIAN BLUEBERRIES >> The rover Opportunity photographed this rippled dune field as it completed a tricky drive into the center of Mars's Endurance Crater in 2004. The photo, made in false color, shows ridges of sand one meter high. The blue tint in the flat sections reflects the presence of small balls of the iron-bearing mineral hematite, named "blueberries" during the mission.

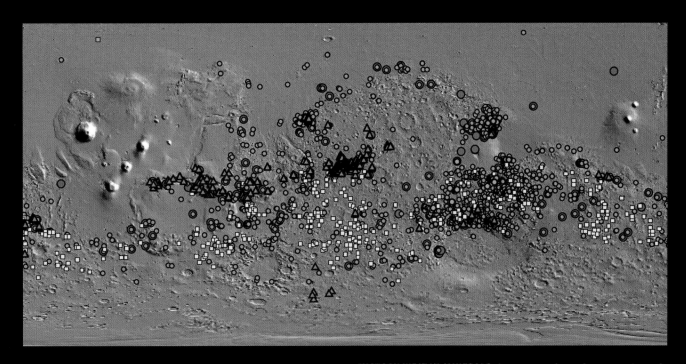

HISTORY WRIT IN MINERALS A map overlay of one region of Mars shows the distribution of telltale formed-in-water minerals. Made by geologist Bethany Ehlmann of Caltech, the map shows clay in blue, carbonates in green, sulfates in red, and chlorides in yellow. These hydrated minerals are generally found in the southern hemisphere and not the north, where lava or another material has eliminated the more ancient geological history.

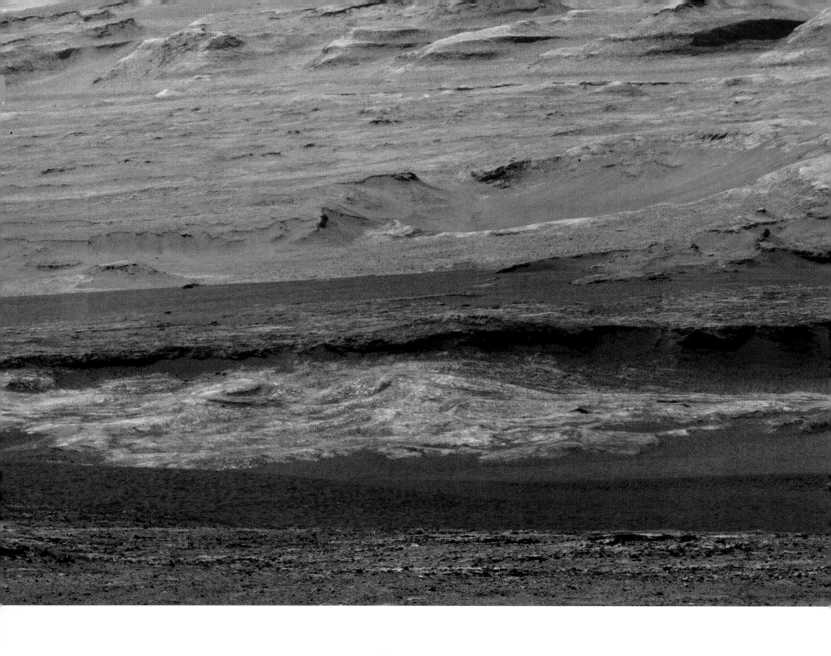

Mount Sharp from four miles
(seven kilometers) away
shows countless layers of
sediment, many of them
filled with telltale minerals.
At the bottom are the black
dunes, farther up a series of
ridges and major outcrops
as well. It's a geologist's
dream for studying the history of Mars.

<< Sunset at the Cliffs of
Moher on the west coast of
Ireland. The layers of sedimentary rock are similar to
those at Mount Sharp and
tell the story of Earth's
geological history, as those
on Mount Sharp tell Mars's.

would ultimately settle in water or form layers that, under pressure, solidified into minerals and other sedimentary rocks.

>> **Small bumps or** nodules cover many of the mudstones in Yellowknife. One theory suggested they were created by ancient gas bubbles, and their presence was enticing to all, but especially to the SAM team, searching for organics. The origins of the bumps, however, remained a mystery.

>> **The hunt for** Martian organic compounds had produced no breakthrough—yet. But the SAM team had several promising leads despite major obstacles presented by the MTBSTFA contaminant and the "perchlorate wall." Painstakingly, the team had learned how to work around the contaminant that had leaked into the instrument and was finding pathways around or over the organic-killing perchlorate. The SAM organics team was increasingly optimistic.

>> **Most important, Curiosity** scientists concluded that the area around Yellowknife Bay had once been habitable. The presence of clays and sulfate minerals formed in neutral water, plus the presence of all the elements needed for life and potential sources of energy, together made the case: The area could have supported life. Never before had such a conclusion been reached for any environment beyond Earth.

Lauren Edgar (left) and Katie Stack (right) are among five Caltech graduate students on the Curiosity team. They've been active in analyzing the layering at Gale Crater, were involved in the original Gale mapping project, and have played key roles in daily operations.

LAYERS OF POSSIBILITY

The findings and interpretations about layering in Gale, and especially around Yellowknife Bay, illustrate both the breadth of discovery so far and the amount that remains unknown. Because as Curiosity geologists, geochemists, geomorphologists, sedimentologists, and more struggled to put it all together, it often seemed as if they were working to put together a puzzle with an unknown but large number of pieces missing.

Exhibit A for this state of affairs is the effort to understand the stratigraphy—or layering— they encountered as Curiosity descended from Bradbury Rise to Yellowknife Bay and then beyond.

Like the stratigraphy of a well-layered wall in the Grand Canyon or what lay ahead at Mount Sharp, the path down featured layered outcrops that were interpreted to represent different periods of time. In classic stratigraphy, the layer on top is the youngest, the one on the bottom the oldest.

Katie Stack, a graduate student with Grotzinger, mapped out the visible layers going down—Bradbury, Bathurst, Rocknest, Shaler, Point Lake, Gillespie, and then the Sheepbed unit in Yellowknife Bay, where Curiosity found mudstones, clays, and a habitable place. From top to bottom, the decline was almost 60 feet (18 meters).

Each of those layers had a particular set of characteristics—fine grained or coarser grained, light colored or dark, vuggy (filled with left-behind holes where mineral crystals once resided) or ropy. Conceivably, each of these layers continued on for miles in each

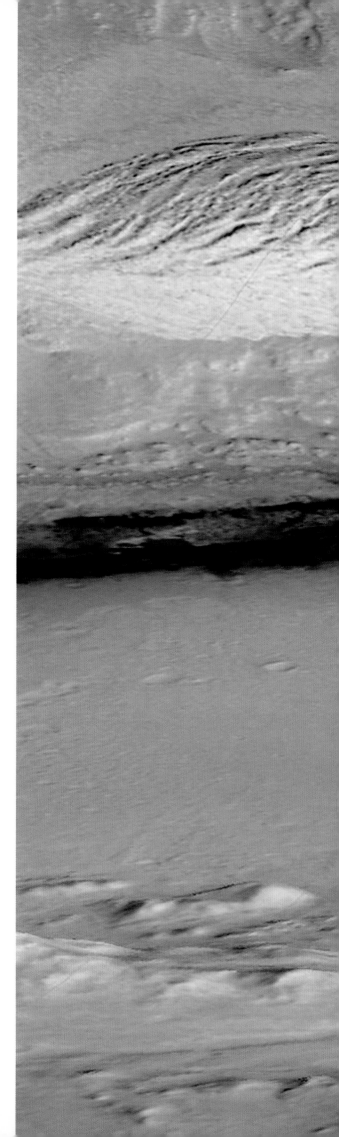

direction, and the exposure at Yellowknife was just a cut into the wedding cake.

The logic of this scenario is that the mudstones at Yellowknife are the end, or near the end, of the Peace Vallis fan. Water from that river, or perhaps an earlier fossil river beneath it, brought all the sediment down from the crater rim. Over the eons it manifested as different layers with different characteristics, depending on the environmental conditions at the time.

Or it could be a far more topsy-turvy story, with the layers continuing and disappearing not only across the landscape but also down into the bedrock. Another theory says, for instance, that the Yellowknife layer has nothing to do with the Peace Vallis fan, and that it is instead connected to the base of Mount Sharp. Yellowknife is lower than the lowest visible layers of the big mountain, but geology can play all kinds of tricks.

"This debate over Gale stratigraphy is a daily hallway kind of thing," said Grotzinger. "It's a very big deal because how it turns out could change the way we understand some pretty basic events and processes in the history of Mars. Depending on which way they go, the results could lead to some real paradigm shifts."

Why would this be important? Because it potentially says a lot about when, and for how long, Gale Crater contained habitable places. After concluding that Yellowknife Bay was once habitable, determining the longevity of that all-important capability to support life (if it was ever present) became a central focus of the mission.

MILLIONS OF HABITABLE YEARS

Eighteen months into the mission, the results were released and were startling: Yellowknife Bay could easily have been habitable for millions to tens of millions of years, well past the point when Mars was previously considered too dry and cold for liquid water.

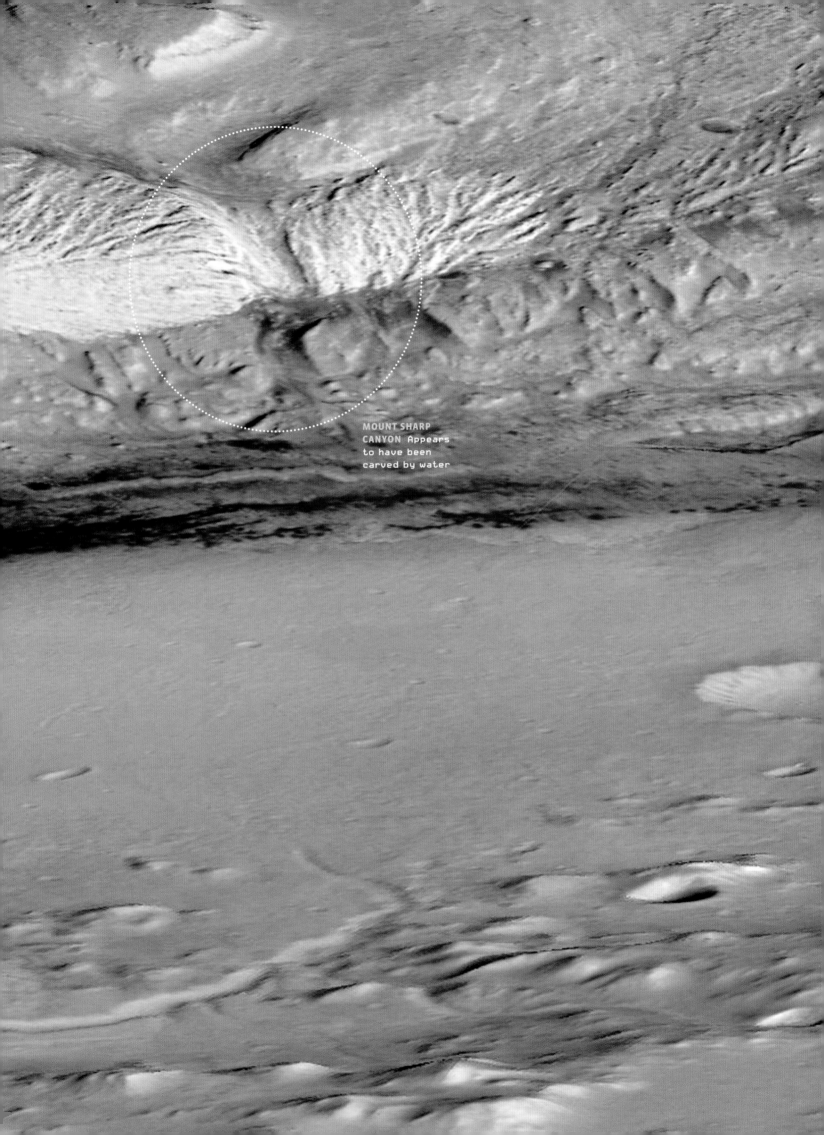

MOUNT SHARP
CANYON Appears
to have been
carved by water

"Maybe Mars didn't become so cold and barren so quickly as many people thought. Maybe it stayed warmer and wetter in many places."

Before Curiosity, Grant was equally involved with the operations and science of Opportunity and Spirit. One of his current interests involves alluvial and delta fans around the Martian globe—some of which may be the latest significant surface flows on the planet associated with snow or rainfall, rather than the more recent "catastrophic floods." He and Air and Space colleague Sharon Wilson Purdy have identified remains of more than 100 of them on the walls of craters like Gale. They have studied in great detail the multiple fans of Margaritifer Terra, which they suggest are among the youngest. Grant says he's been interested in Mars since reading Ray Bradbury's *The Martian Chronicles* as a child.

John Grant, Smithsonian National Air and Space Museum

A geologist and expert in age dating on Mars, Grant helped select Gale Crater as the landing site for the rover, expecting to find interesting formations such as the yardang (below), a streamlined rock carved by wind and sand.

"When it became clear that's where the story was headed, we had a 'Holy crap' kind of moment," said John Grant, a geologist with the Smithsonian National Air and Space Museum and a long-term planner on the Curiosity team. Grant has been studying Mars for decades, poring over orbital images and surface data from rovers and landers. One of his specialties is the very difficult one of setting dates for the ages of craters, valleys, and other features on Mars.

"Maybe Mars didn't become so cold and barren so quickly as many people thought. Maybe it stayed warmer and wetter in many places. Or maybe it was local. But however it ultimately turns out, this conclusion will require some pretty significant changes in how many people think about Mars."

The smoking gun was the finding that those clays contained in the blue-gray mudstones drilled at Yellowknife Bay were not transported there by flowing Peace Vallis water or wind or anything else. A broad consensus within the team held instead that the clays in the Sheepbed mudstone were formed right there at Yellowknife Bay.

What this meant is that well past the time when Mars was supposed to be able to support such things, clay minerals were being formed in very neutral, even drinkable water, at a site that also had other necessary conditions to support life.

In other words, Mars—or at least Gale Crater—was potentially habitable during a time later than imagined possible. The meteorite that formed the crater landed some 3.6 billion years ago, which was already at the endpoint of what has been considered the warmer and wetter era. Then a lot had to happen at Gale before the clays were laid down as part of a habitable environment that the rock record was saying may well have lasted many millions of years.

The story was pieced together from a broad array of less than earth-shattering observations. For instance, the geochemistry of the clay—the information embedded in the chemistry of the samples—told one story. The texture, size, and composition of the tiny rock grains that made up the mudstone told another. The geological surroundings provided important guideposts and clues. And the rim of Gale Crater—the logical source of sediments, if wind or water had carried them to Yellowknife—remained largely unaltered, meaning that the rock had not been broken into the sediments needed to form clays.

And perhaps most important, the Yellowknife samples with the most clay also had extremely low concentrations of the silicate olivine, the most common mineral found in the omnipresent volcanic basalt of Mars. When olivine breaks down, it is available to form clays. So the presence of Yellowknife clay combined with extremely low levels of usually abundant olivine were together strong evidence that the clay-making process occurred right there where the rover was digging.

Grotzinger expanded: "It took months and months for people to become comfortable with this conclusion. We had worked through the question of Yellowknife being once habitable, but we didn't deal with age. Coming to terms with that was a kind of tender moment for the mission."

As well it might be. "What we did was to open up and extend the window of habitability to a later time," Grotzinger continued. "The older paradigm was that lakes and rivers existed only in the

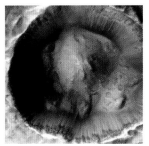

WATER DOWN UNDER Meteorite crash sites reveal secrets. These orbital images suggest there was more subsurface water than water aboveground.

early Mars history, and they were warmed by major impact or hydrothermal processes which would not allow for habitability. Now we see a very different picture."

And this is what that picture of early Gale Crater looks like. It's an incomplete story that doesn't command full consensus, and it brings in theories not addressed by Curiosity findings. But there is some broad agreement: The meteorite impact dug a deep and narrow hole that, over time, grew more shallow and much wider. Water and soil brought in sediments, and the crater began to fill. Whether it filled to the rim is hotly debated, as is whether water or wind was the primary delivery system.

But some time before Peace Vallis began to flow, winds started eroding the crater. This massive excavation left behind Mount Sharp—broad and about 3 miles (5 kilometers) high, and at the center of an otherwise flat crater 95 miles (154 kilometers) in diameter.

Then, millions of years later, rivers and streams like Peace Vallis returned, flowing into the crater and perhaps up to the mountainside some 18 miles (30 kilometers) from the rim.

There is some disagreement about when in this drama the clays were formed—perhaps when sediments and water first filled Gale, or perhaps when water flowed into an already excavated crater. But whenever it occurred, the process of diagenesis—making rock and new minerals out of the deposited sediments—transformed loose material into the mudstone with clays. Later, after the mudstone had dried and cracked, yet another wet period arrived. This time the cracks or fractures filled with minerals—especially calcium sulfate (gypsum) and closely related compounds.

In a paper in *Science* presenting the consensus of the team, Grotzinger wrote that it is entirely plausible that this watery, habitable phase at Yellowknife lasted for millions to tens of millions of years. At landing, hardly anyone would have predicted that.

This breakthrough effort to understand the origin of those clays swept in many of the specialists on the team. They included those who studied sediments and river flows, those who measured and analyzed the makeup and chemistry of rocks and sands, those who could read past environments from current landscapes, and those who could draw conclusions based on the texture, density, and grain size of rocks and sand.

And then there were those, like John Grant, who try to date Martian features by counting craters. Geologists on Earth can establish ages by studying fossils in rock deposits and determining which rocks had the fossils known to be alive in older times and which in younger times. Geologists also use radioactive dating methods on Earth that can give relatively precise readings on the age of rock formations. That technique is only now becoming useful on Mars.

But the most widely used method for age dating on Mars involves crater counts. By counting the number and size of craters in a particular area, Grant can know roughly when in Martian geological history the region was formed.

Areas with many and larger craters were formed during the older (Noachian) era, and areas with progressively smaller and fewer craters are interpreted to be younger (from the Hesperian and Amazonian eras). This is not a function of where the craters are; it is assumed that Mars once had uniformly large and prevalent craters. But later events—most of them involving volcanoes, lava, and erosion—erased big craters in some areas but left them in others.

>> A collection of relatively recent crater hits on Mars, with kicked-up rock, called ejecta, coloring the ground around them. Planetary geologists use crater counting—comparing the number and size of craters in different areas—to make rough estimates of the relative ages of a region and its features.

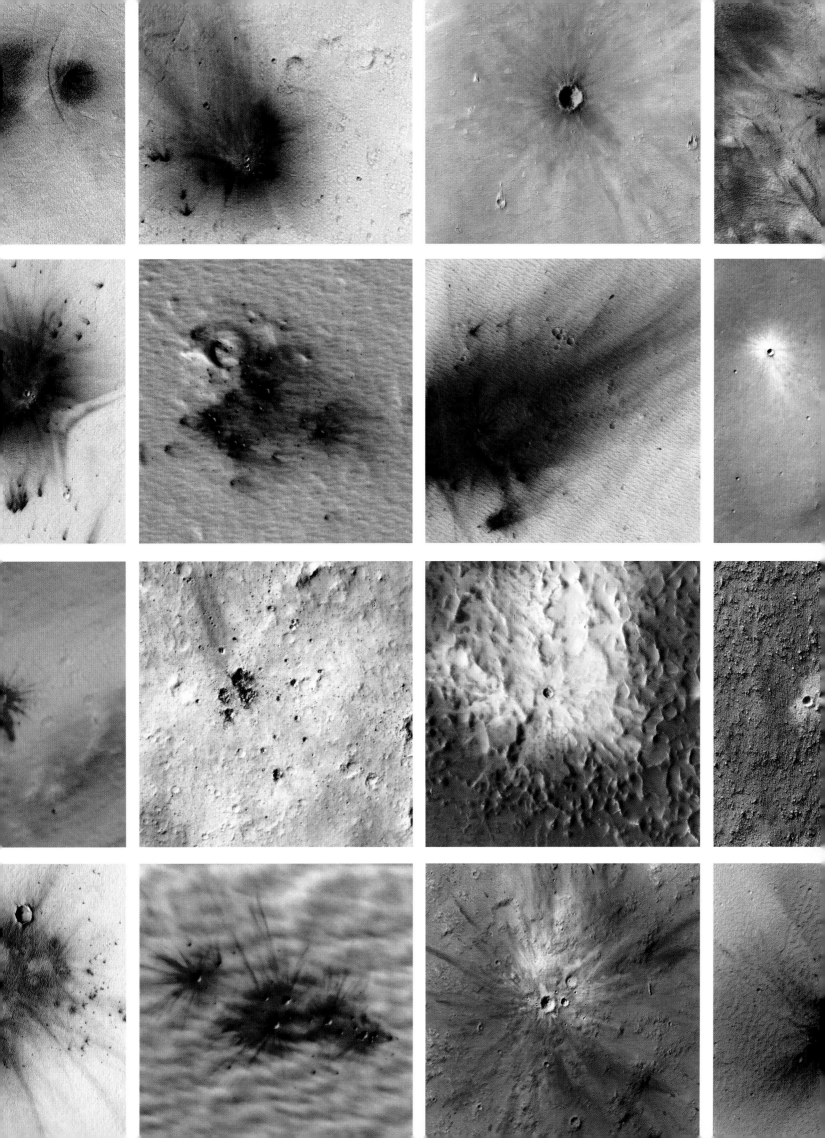

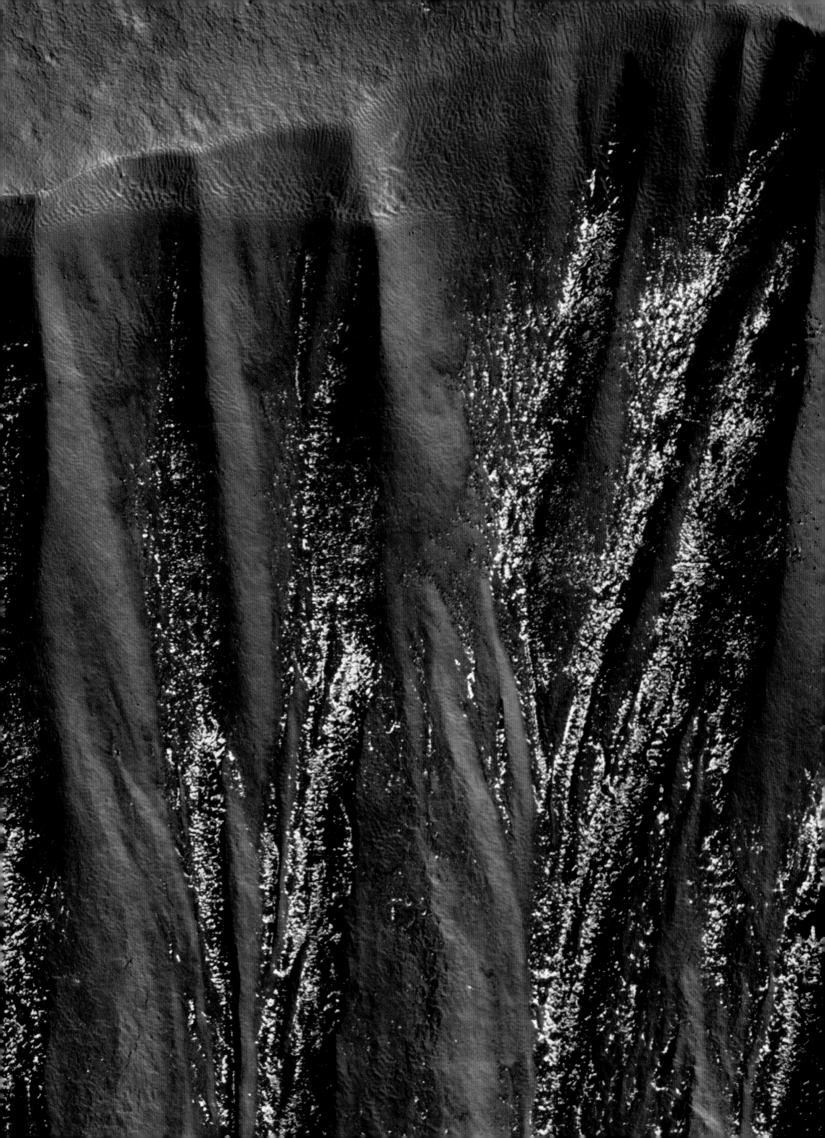

Crater counting is an imprecise method for determining geological age, and the precision shrinks ever more when the areas studied become smaller. But it has been considered, until very recently, to be the best technique around.

Based largely on crater counting, Grant sees Yellowknife as having turned into mineral and rock relatively early in the evolution of Gale Crater (which would nonetheless still be considered late in Martian history). Yellowknife would have been filled with water flowing in from the crater rim, but the sources would have been riverine ancestors of Peace Vallis.

Other Curiosity scientists, such as Dietrich, see Yellowknife as among the youngest features in Gale—the topographically logical end to the Peace Vallis fan. He sees no hard evidence for any explanation other than that. If Peace Vallis is a relative latecomer to Gale, then under this theory Yellowknife and its clays would be relatively young as well, with all the implications about long-term habitability that hypothesis brings.

Some go further. They argue that water, probably from rain or melted snow and ice, continued to flow through the Peace Vallis system past the three-billions-years-ago mark—in other words, into the Amazonian era traditionally characterized by parched and cold conditions such as those on Mars today.

Based on age dating at Gale and also at more than 100 craters around Mars with similar alluvial fans, Grant estimates the last gasp of Martian rivers and streams probably occurred around 2.5 billion years ago. He contends there wasn't enough water then at Gale to run all the way to the Yellowknife area (some 15 miles, or 25 kilometers, from the crater wall), but the upper fan shows signs of a later flow.

The job of geologists in general, and certainly astrogeologists, is to study rocks, fossil rivers, and larger landforms to better explain a little-understood past. Painstakingly, that story was coming together.

Iron magnesium phyllosilicates layer the floor of the crater named Ismenius Lacus.

<< Gullies identified in many Martian craters, such as this in the northern polar region, suggested that water might still run on Mars. But later research found that the gullies move only in winter, suggesting water flows were not occurring.

LOOKING FOR SURPRISES

Scientists look to rocks to confirm theories, but they are most excited when what they find is a surprise. Curiosity, after all, is formally defined as a mission of discovery, and that means always looking for what is not known or expected.

Discoveries like the presence and meaning of those rounded pebbles and conglomerate rocks near the landing site were very important and significant as ground-truthing, but they weren't really a surprise.

The grayish blue insides of broken Martian rocks and the mudstone, however, were not expected and were unlike anything seen before on Mars. The gas bubbles at Yellowknife were a surprise, too, as was a rock feature that appeared on the radar soon after Curiosity left Yellowknife. One rock in particular seemed to have a thin black coating over a small part of it. Black basalt from ancient volcanoes is common in Gale, but the color identified appeared to be deeper. In addition to a Mastcam image of one stone with a small section colored the deeper black, ChemCam also got a good reading on one and came up with a finding both intriguing and complex.

What ChemCam found was a very thin layer of the element manganese covering the rock. This got people's attention because a rock covered with manganese immediately brings up the thorny issue of desert varnish.

In the American Southwest, as well as other dry areas around Earth, this blackish purple covering is common. American Indians created thousands of drawings on these coatings over the centuries, and it can be found on many rock faces.

What is so thorny about desert varnish is the long debate over whether it is the product of dust, wind, and non-biological forces, or whether the unusual concentrations of manganese are a by-product of microbial life. On Earth, virtually all desert varnish is home to some microbes.

The debate initially took an extraterrestrial turn after some Viking images appeared to show a dark varnish atop some rocks, and similar black coverings have been seen by other Mars landers and rovers. Was this a eureka moment—We found signs of life on Mars!—or was it a predictable result in a dry climate with no connection to life?

"It's intriguing because the manganese is quite concentrated and is only on the thin outer layer," ChemCam principal investigator Roger Wiens said. "This is not typical for rocks generally, though it is typical for rocks with varnish." By Sol 342, several rocks had been identified by ChemCam with high manganese—and one very high—but the rapid transit decision did not allow for stops to run more definitive tests.

Nina Lanza is working with Wiens and ChemCam on the Martian varnish possibility, and she has studied rock varnishes using ChemCam technology on Earth. The similarities between the apparent covering on a rock called Caribou and the varnish-covered rocks analyzed on Earth were striking.

"Where we saw high manganese, the Mastcam photos also showed an especially dark area," she said. That manganese reading was on the order of 100 times greater than the trace amounts typically found in Martian dust.

Desert varnish—seen here at Nevada's Valley of Fire State Park, ornamented with 3,000-year-old Native American petroglyphs—is a blackish veneer on rocks found in the southwestern United States and other warm, arid places on Earth. For decades, scientists have debated whether it is the result of microbial activity.

>> Mars rocks like this one, photographed by the rover Spirit in 2006, may be covered with a varnish, although the chemical composition of this sample was not tested.

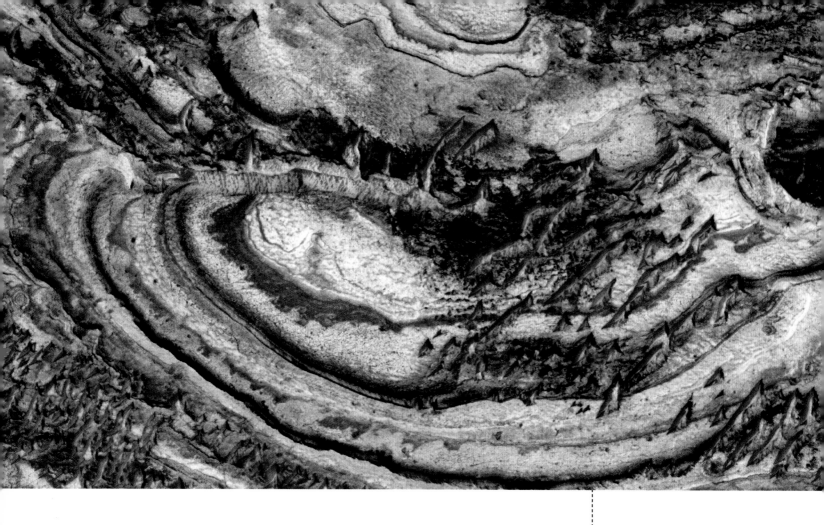

"The dark area was clearly not a shadow," she continued, "but rather features that appeared to be coatings. We need to find many more of these blackened and manganese-rich areas before we can say anything with confidence, but it certainly got our attention."

Desert varnish on Earth is usually found in far more protected areas than the plains of Gale Crater, and some Curiosity scientists are skeptical about the ChemCam observation for that and other reasons. Doug Ming of NASA's Johnson Space Center, for instance, is a soil expert and has seen similar concentrations and black coverings on Mars that he interprets as being formed by a chemical process unrelated to any varnishes.

"They have a hypothesis and that's fine," he said of the ChemCam team. "But I think other explanations are equally or more plausible."

Nonetheless, as Ming has often cautioned, it's wise and useful on Mars to expect the unexpected.

MOUNT SHARP OR BUST

The six-mile (ten-kilometer) trek from Yellowknife to the base of Mount Sharp was never expected to present major science targets. The land was generally flat with small craters and mounds. Most of the visible features were float rocks—millions of large boulders, football- and baseball- and golf-ball-size rocks. The float rocks are generally not local but rather the result of meteorite impacts as far as 180 miles (300 kilometers) away.

The immense power of the impacts, coupled with Martian gravity that is but 38 percent as strong as Earth's, means that rocks shot out of a newly formed crater can travel enormous distances. They can also shoot up into the atmosphere and out into space, where they form

Researchers believe that minerals deposited in this unusual formation near the rim of the giant Valles Marineris canyon include opals and jarosite. The opals are in the bright, relatively white bands, while brown jarosite is concentrated in the brown bands. This false-color HiRISE image shows the layered bands, but it was the CRISM spectrometer on the same satellite that identified the minerals.

^ AUGMENTED REALITY
View this image through NASA's Spacecraft 3D app. Learn more on page 6.

meteorites that may land someday on Earth. (Because our atmosphere is so much thicker and gravity stronger, there are no Earth meteorites landing on Mars.)

Alongside the float rocks were more conglomerates quite similar to those first seen near Bradbury Rise. These new conglomerates peeking out were an unmistakable signal that Curiosity was crossing an area that had also once been covered with water in the form of rivers, streams, or a lake or shallow playa. As ancient-river specialist Becky Williams and others concluded, there was no alternative explanation for how the conglomerates had formed and how those rounded pebbles had come to be deposited.

The Mount Sharp or Bust schedule, however, limited investigation. As Grotzinger and his deputy, Ashwin Vasavada, were keenly aware, some potentially interesting science targets would have to be left unexplored. It's in the nature of any mission and entirely inevitable, if at times frustrating.

And Mount Sharp beckoned.

THE PRESSURE GROWS

With its surrounding buttes, mesas, canyons, and ridges, Mount Sharp promised gorgeous images as well as great science. The journey up the mountain, or at least partially up the mountain, would be a grand showcase for the rover drivers and engineers. But probably most important, the walls of layered minerals and rocks—including the all-important, formed-in-water clays and sulfates—exercised an enormous pull. They represented eons of Martian history written in stone, offering an unprecedented opportunity to read the planet's geological and climatic past.

So both inside and outside the team, the pressure to reach the mountain was growing. At an all-hands science meeting at Caltech around Sol 410, the group supported the idea that one of the five planned science stops on the way to Sharp just might have to be skipped.

This desire to speed up was coming just as the trek to Sharp was becoming increasingly slow and difficult.

Until that point, Gale Crater had enjoyed the frigid but nonetheless relatively moderate temperatures of Martian spring, summer, and fall, but now winter was approaching. That meant the battery that supplied all the rover heaters would be providing ever more power to keep the rover and instruments alive, and less power for science and pedal-to-the-metal driving.

The average drive since Yellowknife had been about 50 meters a day, although the engineers and drivers had found ways to double and triple that rate. But the improvements would have to come during the harshest time of the year. What's more, the flat plains would be giving way in the weeks and months ahead to more rugged and difficult terrain.

So while an especially attractive science target could lead to a detour or even a stop, the entire Curiosity team was eager to reach its prime target and begin the exploration. What's more, it was essential to keep Curiosity in the public eye, and reaching the base of Mount Sharp was the best way to make that happen.

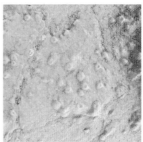

BUBBLES AND CONCRETIONS The MAHLI camera can focus in on details within rocks, including what seem to be gas bubbles (left) and mineral concretions (right).

MARS AND EARTH THROUGH TIME

PRE-NOACHIAN
4.5–4.1 billion years ago (BYA)

- Crust forms
- Some water present
- Thick atmosphere

NOACHIAN
4.1–3.6 BYA

- Water abundant
- pH neutral
- Clay minerals form
- Atmosphere begins to thin

HESPERIAN
3.6–2.8 BYA

- Many volcanoes, sulfuric gas emissions
- Sulfates form
- Progressively more limited and acidic surface water
- Atmosphere thins dramatically

AMAZONIAN
2.8 BYA to present

- Dry and cold
- Very thin atmosphere
- Most water frozen or evaporated
- Iron oxide predominates

4.5 BYA
Formation of planet

4.1–3.8 BYA Late Heavy Bombardment

3.5 BYA
Gale Crater habitable

| 4.5 BYA | 4 BYA | 3.5 BYA | 3 BYA | 2.5 BYA |

4.5 BYA
Formation of planet

4.1–3.8 BYA
Late Heavy Bombardment

3.5 BYA
Primitive photosynthesis

3.8–3.9 BYA
First life-forms

HADEAN
4.5–3.5 BYA

ARCHAEAN
3.5–2.5 BYA

PROTE-ROZOIC
2.5–1 BYA

Proto-Mars and Proto-Earth were not so very different when they formed and began to orbit around the sun, which then gave off about 70 percent of the energy it does today.

MARS

The Amazonian epoch continues to today on Mars. Although the planet has been essentially frozen dry and has only a very thin atmosphere, significant geological changes have continued nonetheless: the formation of Olympus Mons, a huge volcano more than twice the height of Mount Everest; volcanic eruptions and vast lava flows elsewhere; landslides in Valles Marineris, an ever-widening canyon system 4,000 kilometers (2,500 miles) long near the equator; and the development of plains and sand dunes near the poles.

2 BYA — 1.5 BYA — 1 BYA — 500 MYA — 250 MYA — TODAY

2.3 BYA
Oxygen-rich atmosphere

1.5 BYA
First true photosynthesis

1.1 BYA
First multicellular life-forms

500 MYA
Trilobites

400 MYA
Fish

230 MYA
Dinosaurs

200 MYA
Mammals

7.5 MYA
Hominids

EARTH

PHANEROZOIC
1 BYA to present

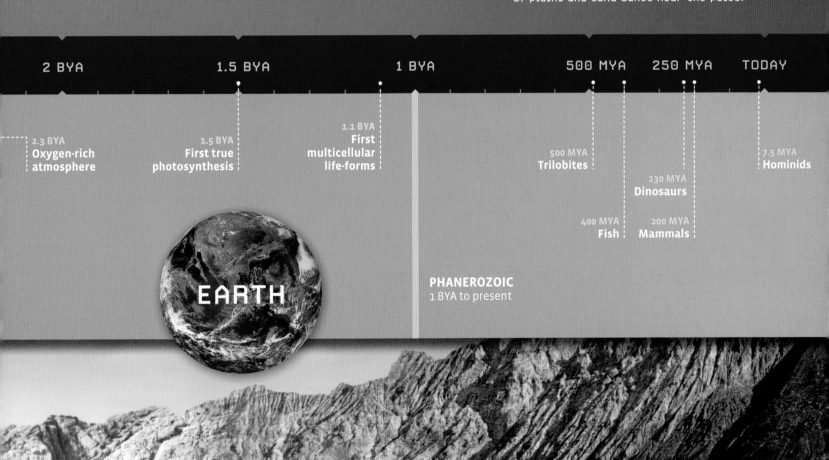

DRIVING CURIOSITY

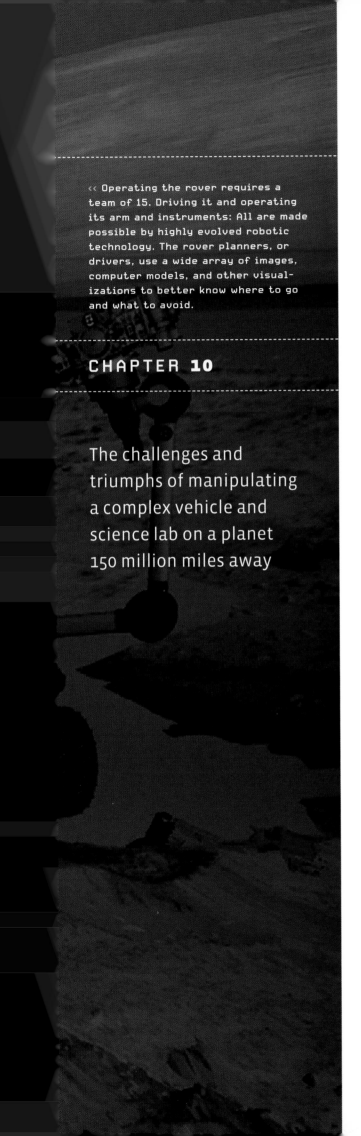

CHAPTER 10

The challenges and triumphs of manipulating a complex vehicle and science lab on a planet 150 million miles away

IT WAS A BIT PAST EIGHT at JPL on Sol 377, and the morning downlink had recently arrived. Vandi Tompkins was the RP1 (rover planner) that morning, which meant that it was her job to map the draft route of the next day's trek for uplink to Mars later that afternoon.

As she knew well, there wasn't much time to take in and understand the exact, excruciatingly precise details of where the rover had ended up after the previous day's drive. And as soon as she did, she had to draw up a plan to move Curiosity safely, quickly, and, she hoped, uneventfully forward.

The first Curiosity team meeting of the day was set for about two hours later, and by then she and her two rover-driver colleagues were expected to have a plausible pathway forward. That plan would be refined, and sometimes totally changed, the day before being translated into a miles-long computer code and sent up to the rover. The Sol 378 drive would be a relatively long one, almost 90 minutes.

But Tompkins and her colleagues wouldn't see the rover in motion; they'd be asleep later that night, or waking up with the expectation (and, on less certain days, the hope) that Curiosity would be following the commands they painstakingly planned, wrote, and sent to Mars.

By nature high-energy and exacting, a robotics expert and mountain climber, Tompkins let her eyes dart from one

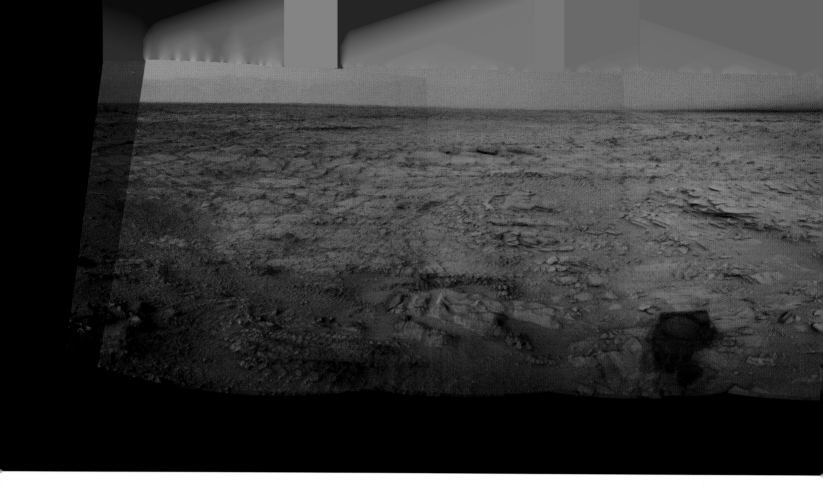

computer screen to another as she pulled up the collection of computer tools that allowed her to place herself in a virtual Gale Crater. A video gamer's delight, but only for the very highly trained. Most Mars rover drivers have advanced degrees in fields like robotics, computer science, and mechanical engineering.

First Tompkins went to the image sent down from the rover's Navcams—the four navigation cameras that show what the rover was seeing relatively close in. That scene was of a seemingly barren and flat desert, with Mount Sharp off to the side.

The goal for the next sol was to drive as far as 300 feet (90 meters) and arrive at a safe site where the rover could do a sample drop. That meant the drive would ideally end on stable

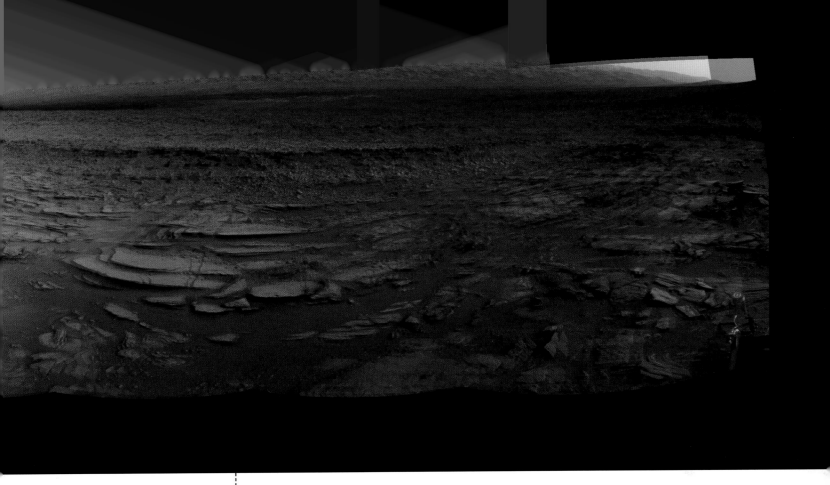

Where the flat image showed a benign patch, the 3-D version showed a number of sandy depressions.

ground, although the length of the drive meant the rover would go past the point where its Navcams and their related computer mesh could see. If all went well, a different set of rover drivers could unstow the long arm the next day and, after an elaborately choreographed set of maneuvers, deliver into the rover body some drilled rock sample from storage at the end of the arm.

Pretty straightforward, as drives on Mars go. But the first impression was deceptive, as it often is. Tompkins reached for her 3-D glasses and saw a rather different terrain. Where the flat image showed a benign patch, the 3-D version showed a number of sandy depressions. Not the kind of places where you want to take the rover; they could be ghost craters filled with soft material that would give the rover a bad day. Both the Spirit and Opportunity rovers had spent some time stuck in unseen sandy holes.

And what about that rock over to the distant right? Was it big enough to cause trouble? The RP1 set out to measure it.

VIRTUALLY THERE

As Tompkins identified hazards and made her plans, she wrote commands that put virtual red arrows on the screen, as well as waypoint darts with white concentric circles on the top. But the main path defining was done with orange lines that simulated each moving wheel.

A virtual rover popped up onto the screen, and soon she was moving it along the first version of her course.

While studying the most lifelike version of the scene, she also glanced frequently at the plain mesh version on the next screen, which underlay the landscape where she was placing her stakes. The animated rover was again on the screen, and in front of it was the landscape as translated into patches of gray computer mesh.

The patches were relatively solid near the rover, but they became more sparse and elongated as the distance from the rover grew. This mesh recorded in triangles of varying distortion where the full imaging machinery had, in effect, determined it was safe to travel. That didn't mean the rover had to stay entirely on the mesh, which after all gave a more precise reading of what lay ahead as the rover moved forward. But it was an invaluable safety tool.

By now, Tompkins's corner in the daily operations room (formally the MSL sequencing room) was a swirl of activity. She zoomed in and out on the screens, slipped on her 3-D goggles and threw them off, turned the images around, up and down.

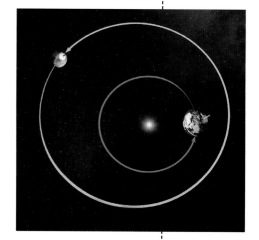

Mars and Earth are so far away that radio signals take from 4 to 22 minutes to travel between them, depending on where the two planets are in their orbits. At the time of Curiosity's landing, the radio signals took 14 minutes to travel one way.

"Hmmm . . . I'm not sure if I need to skirt that depression," she said to nobody in particular. "Do we drive through it or around it?

"I see a little bit of a ridge there. I think we want to stop before the ridge, because we can't see beyond that."

Rover drivers are wont to say that being at their screens is like being on Mars. The 3-D pulls them into the landscape, making the big dark glasses an essential tool. Being able to see the rises and falls of the terrain—the actual conditions the rover will face—is magical for sure, but it's no plaything. It's what allows the human eye to become a full partner with all the other navigation technology.

The intensity of a difficult passage, the necessary focus it requires, blocks out the office chatter around the drivers. Not always, but when it happens it certainly is grand.

Next Tompkins brought up the big overhead picture provided by the orbiting satellites. With surprising detail, it showed the rover in its larger setting and gave a different view of what lay ahead. If an area looked as if it might be a sudden rise or fall, this imaging from above could, in effect, be turned on its side to see the topography of the land. It was a different kind of 3-D, but with the same underlying logic: Take two images of a spot from slightly different positions, put them together, and you have a stereo—or 3-D—effect.

Tompkins wanted to see what the depressions ahead looked like, so she turned the orbital imaging on its side for a look. The pixel resolution wasn't great, but it told the tale and could be exaggerated to tell more.

Her first draft was a modified dogleg that would include a slow turn to keep away from the potentially hazardous depressions. She staked it out, and a set of parallel red lines showed up on the screen to define the drive. A nearby orange line defined what the team called the Rapid Transit Route, the path to Sharp designed primarily to be speedy and safe, but also with those four way stations along the way for investigation.

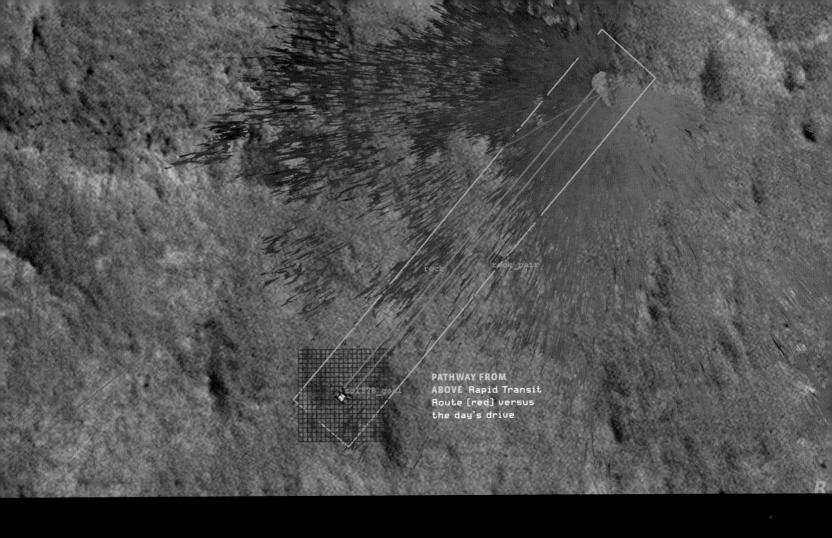

PATHWAY FROM
ABOVE Rapid Transit
Route (red) versus
the day's drive

ROVER SIMULATIONS >> Orbital images are merged with a computer
mesh carrying information about the terrain (above), providing one
of several programs that rover drivers use in their travel planning. The colorful mesh version of the software (below) is used when
Curiosity is driving "autonomously"—without human commands.

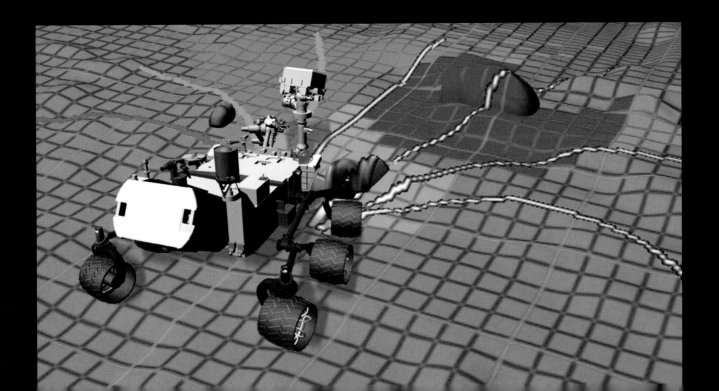

"We don't follow a checklist. We get presented with a new problem every day, and it's fun to be creative about solving them."

NASA has sent four rovers to Mars, and Cooper has driven all of them. He began with the toy-size Sojourner, followed with the long-lived Opportunity and Spirit, and finally Curiosity. He's a manager now, but he says he still gets the most satisfaction out of working with the computer mesh terrains, the 3-D landscape, and the ever present obstacles that make rover driving challenging and appealing. His motto is "Keep it simple," hard-earned advice he takes to heart and shares with the other drivers. Cooper is a former U.S. Air Force officer trained as a computer and electrical engineer, with a JPL specialty in robotics.

Brian Cooper, Jet Propulsion Laboratory, NASA

Cooper is team leader of the 15 rover planners, or rover drivers. The first NASA rover driver for Sojourner in 1997, Cooper helped create the RSVP software that rover drivers use to simulate terrain to be crossed.

The path chosen by Tompkins took Curiosity off the route a bit, but that was fine and happened all the time. Course corrections could always come later.

In rover-driver parlance, this is all called "blind driving"—but the term can be deceptive.

It is not the men and women at JPL who are driving blind—they have all the computer assistance imaginable through the system—called the Robot Sequencing and Visualization Program, or RSVP—that they're created over the years. Rather, it is the rover itself that is driving blind, in that its autonomous navigation system is off and it's dependent on commands from Pasadena.

Auto-nav—whereby the rover periodically photographs the landscape ahead and makes its own decisions about how to avoid hazards and end up where it's supposed to—had been long in coming to Curiosity. The first full auto-nav run had actually taken place just a few sols before, quite a long time into the mission. But it was really only after leaving Yellowknife Bay and pushing hard to Mount Sharp that auto-nav was really needed. The best the rover had done while driving blind was about 325 feet (100 meters) a day. With auto-nav, it could probably reach almost 500 feet (150 meters).

That first major auto-nav drive was orchestrated by Jeff Biesiadecki, who happened to be the Rover Planner 2 for the day. RP2s have the job of completely understanding the drive that's being proposed, the landscape, and the short-term goals for the rover. While the RP1 is the creator, the RP2 is the editor and has full responsibility for the plan.

So while Tompkins was still evaluating the dogleg route, Biesiadecki slid his chair over behind her and made a suggestion. Why not make a straight drive and avoid any turns? A safe path appeared to be available, and it potentially asked less of the rover.

Biesiadecki is low-key and pensive in comparison with Tompkins's visible intensity, but the two nonetheless work together comfortably and quickly came up with a plan B. Sure, why not plan a straight drive if there were no hazards looming?

Tompkins returned to the screens for her due diligence, moving the flags, straightening the pathway, and scanning for potential dangers. Rover drivers have to get along because they work in such close, intense quarters, and—except for those specially trained in controlling the robotic arm—they switch off as RP1 and RP2 all the time.

AN OUNCE OF PREPARATION

The only one with a significantly different role is Brian Cooper, the team leader and a veteran of rover driving and software designing going back to the first NASA rover—the Sojourner back in 1997. Cooper is a manager, but he says the real appeal of the job remains the drives. He was on duty for Sol 377 planning too, and was working the screens and following the planning as intently as the two others. But he would only infrequently chime in, often with guidance along the lines of "Keep it simple. Don't get complex unless it's essential."

For each day's drive, the team gives three extensive walk-throughs describing their plans to each other as well as the full operations team. No other rover team has that kind of load, but no other team could drive the rover off a cliff.

INTRODUCING THE PATH

The first presentation was made by Tompkins and showed a straight shot from the previous day's endpoint to a seemingly benign area about 300 feet (90 meters) away. The route as drawn on the Navcam screen was the focus and appeared on monitors around the room, and from around the room came questions about some rocks, about timing, about some science testing planned along the way.

Tompkins then showed the mesh version of the drive, which included some significant areas where there were no gray triangles—meaning there was no confirmed information about the spot. A manager voiced concern. But then Tompkins showed the Navcam screen, as well as a Mastcam screen with more distance and the orbital imaging, too. It wasn't that dangers were lurking where the triangles were missing; there just were few features to be picked up and recorded.

While the number of people in the operations room was a relatively small 25, many others were listening in by phone. As the first meeting was wrapping up, Robert Sullivan, a Cornell University expert on the Martian surface, chimed in with a thank-you to Biesiadecki and the rover team for a clever maneuver the day before that helped make possible some interesting science.

As Biesiadecki explained, the scientists had asked the rover planners to drive into a particular sand ripple, a hardened wave formation of small-grained sand. It was only a few meters off the path, so the drivers arranged to push a wheel into the ripple and expose for the cameras its insides, which were intriguing to some on the science team.

The effort and results were rather like the numerous times that the rover wheels had both intentionally and unintentionally driven over rocks and broken them in half, exposing scientifically intriguing blue-gray and green insides free of the butterscotch dust covering. The wheels on the Spirit and Opportunity rovers had done similar work as an extra Mars excavation tool, but Curiosity had far greater crushing and exposing capacities because it's so much bigger.

"Great, great work, you guys," Sullivan said over the operations room speaker phone. He said the scientists were all over the images taken after the ripple scuff, and they had proven to be revealing. "Really, it was a pleasure to tell the others that you folks, the rover planners, had designed the drive and made it work," he said. Biesiadecki said he would be looking for other ripples like it to puncture, and Sullivan, a team surface properties scientist, couldn't have been happier. Science on the fly.

CHANGING GEARS

Soon after the operations team meeting, the three rover drivers got together for their own review. Tompkins went through her plan again, though in greater detail, and prepared to make a surprisingly formal handoff to her colleague. When the meeting was completed, the computer file for the entire plan would be moved from Tompkins's name to Biesiadecki's.

But before that happened, Cooper and Biesiadecki proposed a change: The rover would stop at a halfway point to undertake a review of exactly where it was. Using Navcam imaging and the process of visual odometry, which uses sequences of images to estimate distances traveled, the rover can determine if it's off track and make corrections—what's called a "slip check."

A test model of Curiosity was taken to the Dumont Dunes near Death Valley, California, so rover drivers could practice operating it in the days before landing. The test model is called Scarecrow because it doesn't have the computer brains of the real Curiosity.

>> Gray lines represent a computer mesh showing ground determined to be safe to drive on. The rover drivers generally command the rover, but Curiosity has been allowed to drive autonomously more often on the long trek to Mount Sharp.

This is an important process because, even on the relative flatlands where Curiosity was driving, the rover is constantly slipping due to even tiny tilts in the land. Those small slips can become big slips over time, and so this kind of checking is essential. And it will clearly become more of an issue when the rover reaches the dunes on the way to Mount Sharp, the more rugged base of the mountain, and then embarks on the climb itself.

"There's just a limit to how precise we can be, especially on a long drive," said Cooper, who has conducted more than a few. He was not only the original rover driver on Sojourner but also logged time with Opportunity and Spirit, all the while designing new software to better understand and control the driving. "We can and we will keep Curiosity safe," he added, "but there will always be risks out there. Some we'll know, and some we won't until they jump up to bite us."

The rover drivers for Sol 378 left their stations late in the afternoon, confident that the next morning they would get e-mail messages telling them that the rover had indeed driven 300 feet and was now ready now to start up the robotic arm and drop some sample into the chem labs.

A ROCKY ROAD

But it didn't happen that way. Instead, the next morning's downlink showed that Curiosity had driven 300 feet and was in the right area, but its left rear wheel was perched on a rock.

Not a big rock—only about two inches (five to six centimeters) in diameter. But it was under the edge of the wheel, and that had greater significance for balancing, or unbalancing, the rover. The rover arm makes up about 10 percent of the weight of the rover, and unstowed it can slightly change the center of gravity, a worrisome possibility.

The operations room was buzzing after the news came down, as experts of all sorts weighed in about the risk posed by the small but awkwardly placed rock.

The main concern was not that the rock would unbalance the rover and tip it over if the arm was extended. It was rather that if the wheel slipped off the rock while the arm was out, the shaking and vibrations from even a minor fall could potentially damage the arm or propel it into the rover mast. A ton of rover could certainly keep a small rock in place; that was certain. The extreme temperature changes between day and night added another hazard to the mix: The resulting thermal expansion and contraction could cause Curiosity to inch itself forward or backward, making its location atop the rock ever more uncertain and potentially hazardous.

Curiosity managers and engineers have been consistently conservative when faced with risks or untried maneuvers. The mission had been significantly slowed down as a result, but almost all of its parts were running fine—a not-inconsiderable achievement after substantial time on Mars. So nobody was surprised when the team managers decided against unstowing the arm until the rover could be moved again the next sol—this time a shorter and more predictable 50 feet (15 meters).

VISUAL ODOMETRY Using the Morse code signature made as the rover's wheels turn, drivers have a way to measure how far it has traveled.

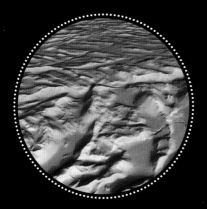 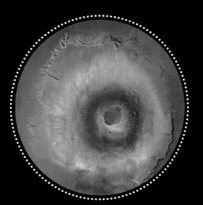 

DETAILS FROM ON HIGH >> As the surface of Mars becomes ever more visible from above, details become both useful for driving the rover and scientifically important for understanding the planet. Landslides are visible around the rim of a large crater (above right), with piles of rock on the crater floor. At 16 miles high, Mars's Olympus Mons (center, image color-coded for height) is the tallest volcano in the solar system. The parallel ridges and valleys of Sulci Gordii (left) tell of a major collapse in the flank of Olympus in the ancient past. All three images were taken from the European Space Agency's Mars Express satellite.

BUMP IN THE ROAD The drive on Sol 379 ended unexpectedly, thanks to a small rock. Although the rock was hardly large enough to harm the rover, it reduced its stability enough to make the rover drivers cautious about using the robotic arm. However much technology and care go into planning a drive, a rock under the wheel is sometimes unavoidable.

Data from previous missions, which provide a
wealth of information when layered together, help
Curiosity drivers understand the terrain ahead in
terms of slopes, obstacles, and contours, but the
resolution is not high enough to eliminate the need
for constant analysis from the rover itself via the
Navcam and Hazcam engineering cameras.

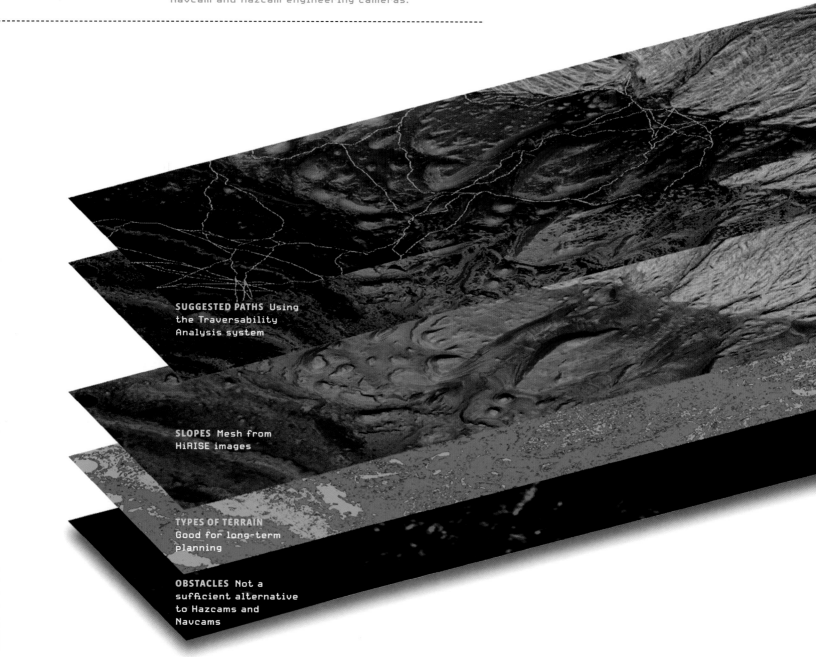

SUGGESTED PATHS Using
the Traversability
Analysis system

SLOPES Mesh from
HiRISE images

TYPES OF TERRAIN
Good for long-term
planning

OBSTACLES Not a
sufficient alternative
to Hazcams and
Navcams

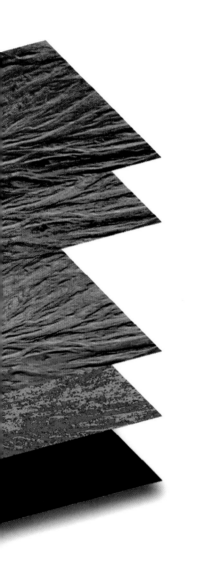

The rock had cost them a day. As the Curiosity team found on "anomaly night," there are plenty of other snipers out there that could end the mission with one on-target shot, so every day is precious. Still, this was an unavoidable glitch, one of many that arise no matter how careful drivers and others on the Curiosity team might be.

"Mars is always throwing up surprises, challenges," Cooper said the next morning.

After the rover has been through more situations like this, he says, mission managers will surely be more inclined to go ahead with an arm maneuver despite the presence of a rock under the tire. It's a question of understanding the risk, then getting comfortable with it.

"So the lesson is just to roll with the punches, respond to whatever comes up," he said. "To be honest, that's part of what makes the job so much fun—the variations. Yeah, we lost a day, but we had to think hard about the situation we were in and find the right way out."

On Sol 378, rather than having the arm deposit one of the samples of drilled Yellowknife rock into the rover, Curiosity was moved off what most agreed was a minor if unignorable hazard and driven a short distance to clearly flat terrain with no rocks in sight. And on Sol 379, the arm was unstowed and the sample drop maneuver began.

THEORY MEETS PRACTICE

In theory, the task of robotically transferring a baby aspirin–size sample of pulverized Mars rock into a small receiving hole does not seem that difficult. Robotic manufacturing on Earth has made that kind of maneuver pretty common and unremarkable.

But on Mars, the transfer requires somewhere on the order of 600 commands. It can take 30 minutes or more. And while it ranks as one of the more straightforward tasks undertaken by Curiosity's seven-foot robotic arm, it nonetheless involves risks that could seriously harm the mission.

Matt Robinson is the lead of the robotic-arm rover planners—the arm drivers—and his job is to make sure that the five-jointed arm does what it's supposed to.

That means enabling the hard-driving drill to dig into rock and then deliver the results over several operations to the rover chem labs, maneuvering the hand lens camera and one of the instruments designed to make contact with Mars to within inches of their targets without causing a crash, and figuring out how to do this and more when the rover is at a 20- and potentially even 30-degree incline.

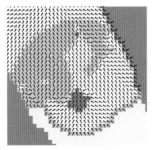

WARNING SIGNS Before rover planners command the rover to drive autonomously, they map out red "no-drive zones."

And like his rover-driver colleagues, he does it all through day-before programming and sends the commands an average of 150 million miles away to Mars. When operating the arm, he also has to deal with the fact that gravity on Mars is only 38 percent as strong as it is on Earth, creating a kind of thermodynamic fun house. The rover has fault-protection controls to stop the arm from doing many bad things, but still, driving the robot arm is not for the faint of heart.

Robinson showed up to plan the Sol 379 sample drop aware that something was amiss. When you're that engaged with an

It's such a part of him that he can move his
own arm with the same almost tai chi movement
as Curiosity's arm at work.

endeavor like Curiosity, and when information about your role in the mission is routinely there in your e-mail, it's generally impossible when you're home to ignore what the rover is doing.

Several teams were already at work: the soil-properties scientists trying to understand the finest details of the terrain, specialists in understanding the dynamics of when and how the rover will slip, and the engineers who would have to work out the actual sequences that day.

The tactical uplink lead is the person who runs the daily operations show, and he was collecting information about how the team might be able to go forward with the sample drop. But nearby was a mission manager, whose job is to keep the rover safe. As Robinson put it, "the uplink lead generally plays offense, and the mission manager is there to play defense." Pushing forward versus health and safety.

Robinson went back and forth to the different groups, and within an hour they had concluded that there was indeed sufficient risk to warrant a delay. They were being conservative, but Robinson says it was part of a larger game plan, one that involved the inevitable risks ahead on the slopes of Mount Sharp.

"Our underlying thinking is that we don't take risks unless the science warrants it," he said. "If it looks like we might get some great result, hit a home run by pushing the arm and rover, then we'll do it. If not, then why ask for trouble?"

Like Brian Cooper, like computer-anomaly solver Magdy Bareh, like drill maker Louise Jandura, Robinson is so good at what he does in part because he has been part of the team that designed the system he operates. He's lived it for years—simulated it on the screen, worked with it on the test bed, planned the movement sequences, had it verified and double-checked, and essentially knows it inside out.

It's such a part of him that he can move his own arm to the same positions and with the same almost tai chi movement as Curiosity's arm at work.

That's not easy, given that the arm has five sockets—what engineers call "five degrees of freedom"—and so every arm joint has to move on cue. But by heart, he can now deliver a sample from the collection device at the end of the arm to the tiny inlet funnel on the body— the maneuver that takes more than 600 commands.

And that isn't just a party trick. Rover and arm drivers have a vast amount of technology at their disposal and can know to the finest detail the conditions they face, the location of the rover and its arm, the vehicle tilt, and the arm stability. And they get a near constant stream of images to let them understand what changes with their commands. But still.

"There is an undeniable element of instinct involved, of knowledge that comes with years and years of knowing how it all works," said Robinson, who also led the robotic-arm team on the Phoenix mission, which found water ice below the polar surface. "It's like the mechanic who works on your car. Someone else might know as much about cars, but if that

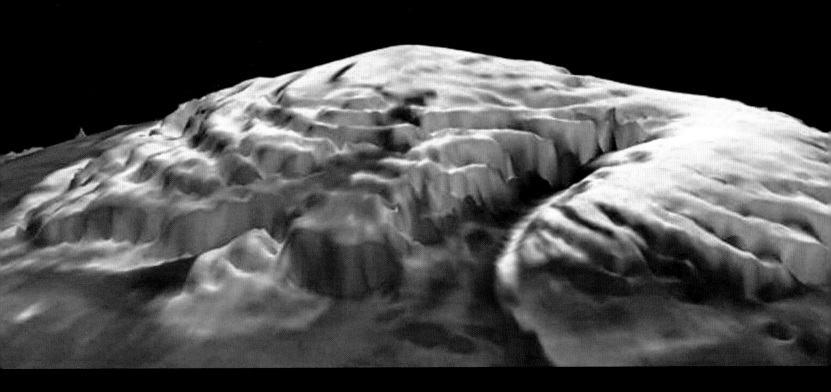

MAPPING FROM ORBIT >> Using information from the Mars Orbiter Laser Altimeter (MOLA), NASA researchers created this graphic of elevations at the planet's north pole. MOLA flashed laser pulses toward the Martian surface from the Global Surveyor spacecraft and recorded the time it took to detect the reflection, then translated the timing data into a detailed topographic map.

STUDENT DRIVER Wheel tracks after Curiosity's first autonomous navigation drive illustrate that the rover chose a path to avoid a small pile of rocks.

Waypoint1

mechanic has experience on yours, then he understands the nuances, the stuff that isn't in the manual.

"I just know what will really happen when the arm is moving around—if it will shake, if it will buck, what it will do under different conditions. The other drivers know it too. It's not like we have to model the maneuvers, because over time we can actually visualize the flight software."

REAL-WORLD CHALLENGES

The big tests for the rover and arm drivers lay ahead in the canyons and on the slopes of Mount Sharp. The first soil scoop and first drills were intentionally on flat ground as a kind of practice run—though they turned out to hit the jackpot.

But for as much as a decade to come (if the rover holds up as hoped), the terrain will get progressively more difficult. The goal, after all, is not only to intensively explore the scientifically rich base of the mountain, which at some points is very difficult to access. The rover will also climb up Mount Sharp as far as it can safely go.

While Curiosity was cleared to drive on inclines up to 30 degrees by this point, the rover team was allowed to perform contact science—drilling, touching rocks, brushing them clean of dust—only on angles up to 10 degrees. For the Mount Sharp expedition, that wouldn't be enough.

That's why the JPL Mars Yard—with its simulated Martian hillside and large array of soil and rock types—remained busy five seasons after landing, with engineers and rover drivers measuring rover slip at different angles, analyzing how the drill, with its percussive banging, behaved on slopes, and generally learning what was and wasn't possible on increasingly steep

After learning from their software and imaging the best way forward, drivers plot out their path using virtual pegs in the ground.

‹‹ Preparing for what is called contact science—the rover's working interaction with the surface of Mars—requires planners to draw up a plan for the use of instruments on the turret by arraying the same virtual pegs, lines, and arrows that later become computer commands sent to Mars.

terrain. To be safe, computer engineers were also creating new fault protections for the rover so its computer would automatically stop everything if an activity on an incline had become dangerous, or if the arm was getting too close to the mast for comfort.

As for the drivers, their own challenges would clearly grow, too. Imagine the difference between driving over relatively flat land with occasional sand, crater, and rock hazards, and then taking on dune fields, scrambling over rocky terrain, snaking through pathways between rock faces. Or reaching the arm full up at awkward angles to sample bits of Mars that just might hold clues about ancient conditions and even ancient life.

Driving the rover is actually very little like playing a video game; it involves way too much responsibility, precision, and consequence for that. But there is an undeniable adrenaline rush, and that's why once you're a rover driver you tend to remain a rover driver. Of the 15 rover drivers on duty when Curiosity landed, 14 were still there for the push to Mount Sharp.

TRAVERSE
OF CURIOSITY

137°25'E

N

4°36'S

METERS
0   100   200   300   400

0   400   800   1200   1600
FEET

MAP KEY
Landing site
383☐ Way point and mission day (sol number)
Traverse path
Projected route to Aeolis Mons (Mount Sharp)
Descent blast zone

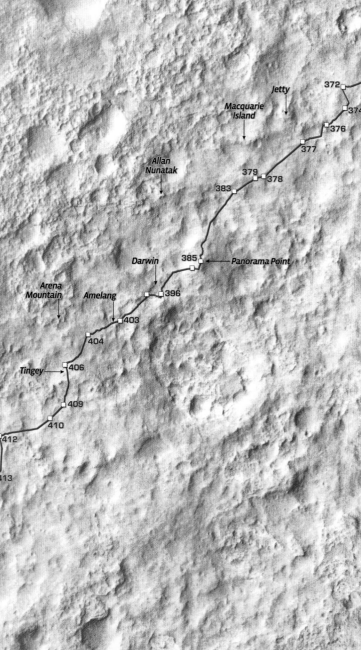

Jetty   372
Macquarie   374
Island   376
                                 377
Allan   379 378
Nunatak   383

Darwin   385   Panorama Point
Arena   396
Mountain *Amelang* 403
              404
Tingey   406
            409
Weaver   410
            412
Slide
Mountain   413
Beers Hill
417

4°37'S

419

422

424
426
429
431
433

436
437
438
454   440
455   453   → *Cooperstown*

137°26'E

137°27'E

BRADBURY
LANDING

Goulburn

Link

*

26

Coronation

29

39

41

Hottah

Jake
Matijevic

45

49

John Klein and
Cumberland

Point Lake
Rocknest
Bathurst

297

YELLOWKNIFE
BAY

307

Shaler

GLENELG

335

338

340

342

343

344

345

Elsie
Mountain

Yellorex

Mount Wilson

Bell River

Twin Cairns Island

349

351

354

356

358

Mealy
Mountain

361

363

365

369

370

Clarabelle

Mount
Berg

Kennedy
Mountain

TREK TO
MOUNT SHARP

AREA
FEATURED

Bradbury Landing

*

Yellowknife Bay

297

Darwin

Cooperstown

455

KILOMETERS

0   1   2

0   1

MILES

N

Mount Sharp Entry Point
(Murray Buttes)

A E O L I S   M O N S

(MOUNT SHARP)

THE PROMISED
LAND

<< Mount Sharp, or more formally Aeolis Mons, is more than 3 miles (4.8 kilometers) tall and about 60 miles (96 kilometers) wide. It has several peaks: What looks like the top of the mountain here is only one of them. A Curiosity Hazcam took this photograph more than five miles from the base of the mountain.

CHAPTER 11

It's three miles high and layered with minerals and rocks that can tell the history of Mars

ITS ORIGINS ARE a mystery and its presence a puzzle. Yet it may just be the guidebook to Martian history that scientists have long been seeking.

It poses the most hazardous, complicated challenge a Mars rover has ever faced, with steep inclines and descents, tucked-away recesses that have to be probed, canyons and buttes where communication with Earth may be limited or impossible. Yet it's the primary reason why Curiosity was sent to Gale Crater.

It has large fossil channels that flowing water may well have cut. Yet the upper half of the mountain shows little sign of ever being touched by water and may well be a huge pile of Martian sand and dust cemented into rock.

Originally and formally, the Curiosity mission plan was to explore the area around the landing ellipse for a limited amount of time and then make a beeline to the antici- pated marvels of this enigma, Mount Sharp. Rising more than 18,000 feet high in the middle of Gale, it features layer upon layer of the kind of exposed Martian rocks and minerals that planetary geologists, geochemists, geomor- phologists, and astrobiologists have only been able to dream about.

Yet after months of exploring at the Yellowknife detour too good to leave, the rover team had to make a dash to the mountain in the midst of the frigid Martian winter just

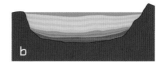

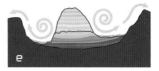

Theorized stages of
Mount Sharp's forma-
tion: A meteor hits,
forming a large crater
[a]. Water and wind
bring in material that
covers the crater [b].
Wind erodes the sedi-
ment out, but it returns
and builds the mountain
higher [c, d]. Wind
scours the crater and
leaves Sharp standing
in its midst [e].

>> Gale Crater was
formed when a meteorite
crashed down some 3.6
billion years ago, dig-
ging out a bowl that is
now more than 90 miles
[145 kilometers] wide.
Impacts like these
create enormous heat
and kick out debris
for hundreds of miles.

to arrive before the end of the rover's primary mission That "primary" lasted one Mars year, or 687 Earth days.

Is it any wonder that to the Curiosity team, the entry pass to the lowlands of Mount Sharp has been universally known as the promised land? It just made the wait harder that the mountain—mesmerizing in its size, its bareness, its secrets—was always in plain sight.

DESTINATION: MOUNT SHARP

As Curiosity and other scientists have found, Mars often behaves in ways that have parallels on Earth. Many features, formations, and processes look and act differently because the planetary histories have diverged, but they are none-theless familiar.

Not Mount Sharp.

There's nothing like it on Earth; nature here just wouldn't oblige. We have cliffs of layered sediments, but they are cut into the planet (think the Grand Canyon) and not piled up on top. Rather like the giant Olympus Mons volcano on Mars—which at 14 miles high is three times the height of Mount Everest—Sharp lets you know that you're on a different planet for sure.

Nonetheless, this much is known: Mount Sharp, or more formally, Aeolis Mons, was formed sometime after a cataclysmic meteorite strike some 3.6 bil-lion years ago.

Given the current size of the crater it sits in, scientists estimate that the meteorite that hit was substantial but not huge, and that the initial crater was seven to ten miles deep. What is called that initial transient crater was relatively narrow and soon developed a peak—a rebound of rock that pushed up from deep on the crater floor. The peak, which was a steaming pile of pul-verized rock, rose about one quarter of the way up the transient crater hole.

Then the steep sides of the crater began to fall in, making Gale much wider (now about 95 miles, or 154 kilometers, in diameter) and consider-ably less deep (rising up to 3 to 5 miles, or 6 to 8 kilometers, below the rim). Scientists know this process because it plays out similarly whenever a large meteorite crashes into Mars or elsewhere around the solar system.

What happened in the eons that followed to form the Mount Sharp of today is hotly debated, in part because the answers have such potentially great scientific significance.

The mountain provides what may well be a window into the contested geological history of the entire planet, and so some existing theories will gain support and others will falter. Mount Sharp certainly won't tell the entire story of Mars, but it can easily force major changes in how the planet is understood.

"Mount Sharp has been studied like no other mountain on Mars, but it's all from orbit," said Ralph Milliken, a geology professor at Brown University and an early advocate for the Gale landing site who has written extensively about it.

"It's hard to understand the stratigraphy of the crater wall, it's hard to define what's actually the base of Sharp, and it's hard to know what's sitting beneath or above what," he said. "So things are tricky, and it's been easy to be fooled."

Now comes the ground-truthing that will begin to clarify those orbital observations and unravel the mysteries.

WHAT WE KNOW ABOUT MOUNT SHARP

That process begins with a general consensus about at least some basic aspects of the Mount Sharp of today.

It's agreed, for instance, that the mound is not an extension of the original crater rebound peak. That Mount Sharp is curved in a wide crescent, with two distinct and separated horns that rival the peak, makes clear that it's something entirely different.

It's agreed that the mountain is made up largely of sedimentary material: rocks, sand, and dust eroded and transported from their original locations by water, wind, ice, other meteorite strikes, or airborne ash from Martian volcanoes. That was nailed through orbital photogeology.

As Curiosity got closer to Mount Sharp, testing intensified for operating on steep inclines. At the JPL Mars Yard test bed, drill cognizant engineer (or "coggie") Avi Okon and sample system chief engineer Louise Jandura worked the drill system on a 20-degree incline.

It's agreed that the base of the mountain is ringed with substantial deposits of minerals that can be formed only in water. Instruments on the same orbiting satellites can read the rocks for signatures of certain minerals, and they came through loud and clear around the base and lower mound of Sharp.

And it's agreed that the top of Sharp is higher than the northern rim of the crater, though lower than the southern rim. This is quite puzzling to those trying to understand how the mountain was formed because most theories involve a long-ago filling of the Gale basin with sediment. Why would a mound keep on growing above some of the crater walls?

What all this means to planetary geologists is that the mountain was formed differently from how mountains rise up on Earth—generally via volcano or the slow but all-powerful process of plate tectonics.

Neither a volcano vent nor lava fields have been identified atop or around Sharp, basically ruling that possibility out. And as for plate tectonics—the mountain-building, landscape-altering, rock-recycling dynamic associated with the movements of massive but separate pieces of the planet's underground floor—it clearly is active and central on Earth, but it has never been convincingly detected on Mars.

Given this strange set of characteristics, how might a team of scientists 150 million miles away go about exploring and trying to understand the place?

The answer involves some inevitably riskier maneuvers, though within strictly controlled guidelines: a preplanned (but changeable) scientific campaign, a push to the places where different mineral types and rock units are known to abut, and a new aggressiveness in using the well-broken-in rover.

They will go to ridges and troughs where orbiting satellites have detected minerals most likely to collect and preserve organics. Although these minerals were ultimately found at Yellowknife Bay, the satellites had not detected them beforehand. The fact that signatures of these minerals have been strong and widespread at the base of the mountain suggests a formed-in-water mineral mother lode.

The terrain has endless hidden caves and outcrops, nooks and crannies where the organics known to arrive from space certainly land. In those hidden places, the incoming organics as well as some hydrocarbons potentially formed on Mars might be protected from the radiation and other forces that are breaking them down on the unprotected plains.

What's more, large deposits of clays are known to have chemical properties conducive to preserving organics much better than other rock and mineral types. The clays in effect attract and cleave to organics more than most other minerals, and they also tend to be found in areas where organics settle in substantial amounts.

So Mount Sharp may or may not turn out to be the promised land for the scientists of Curiosity. But it is almost certainly the most promising site beyond Earth that a human or human-made machine has ever visited.

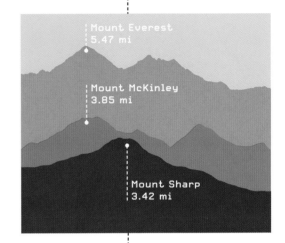

Comparing Mount Sharp to the tallest peaks on Earth: Sharp is just shy of Mount McKinley, and, while about 10,000 feet below Mount Everest in height, in breadth it is more like Everest than the others.

Mount Sharp's vast and varied
sedimentary layering offers Mars
scientists an area unprecedented
for its rich discovery opportunities.
For the engineers and drivers, though,
it presents a challenge like no other.

<< Robert Sharp [1911-2004] is considered
one of the fathers of planetary science.
NASA named its prime Curiosity target
after him, but the International Astronomical Union objected and renamed
the mountain Aeolis Mons. Still, the
destination remains Mount Sharp for
the Curiosity team.

Mount Sharp is named after Robert Sharp, one of the pioneers of planetary geology.
A professor and geology division chair at Caltech for half a century, he was known and admired
as a mentor to many students.

The Gale mound had been identified in the 1970s, but it had no name until well after the decision to land Curiosity nearby. Many of those who benefited from his mentoring lobbied for the name
Mount Sharp, and it was given in 2012. But that actually isn't the correct name of the mountain.

The International Astronomical Union, which is responsible for naming large planetary
features, looks to light and dark qualities of the features that need a name and comes up with
a classical name based on what it sees. Thus the name Aeolis Mons, which refers to an area in
ancient Greece. In recognition of NASA's achievement and in honor of Sharp, the IAU named a

large crater about 160 miles west of Gale after the pioneering professor. But to the Curiosity team, JPL, and NASA, Aeolis Mons remains Mount Sharp.

One of those mentored by Sharp and eager to name the mountain after him was Mike Malin, who along with Ken Edgett is one of the photogeologists who helped revolutionize the study of Mars with their orbital cameras.

They were the first to understand that Gale Crater and the strange mound at its center were important places. That was in a *Science* journal article in 2000, where they made the then controversial case for widespread sedimentary layering across Mars. Gale was one of numerous craters or canyons described with this kind of layering, which appeared to be cleanly exposed for miles on the central mound.

When the NASA site selection committee considered Gale for the Curiosity landing, most of its interest focused on that layering and the most unusual mountain that it graced.

WHAT THE ROVER MIGHT MEET ALONG THE WAY

Fast-forward to the rover as it approached Sharp. Here is some of what the Curiosity team now knows lies ahead:

For starters, the approach to the base of Mount Sharp takes the rover along a dune field of very dark sand, rich in the mineral olivine (the most common mineral on Earth). The dunes are known to move with the winds, which can get quite strong. The sand includes minerals formed in water, but a dune field is not a likely candidate for habitability. It is, however, likely to be a visual delight, with dunes reaching as high as 35 feet (10 meters) alongside the rover path.

What is considered the entry point to the apron of Mount Sharp is a collection of buttes sticking out of the dune field, named during the rover trek for the recently deceased Bruce Murray,

THE CANVAS
Where the
ChemCam pictures
were taken

A STACK OF LAYERS
The prize ahead

A LAYERED LOOK >> Made of 22 images in all, as
seen through the ChemCam fish-eye lens, this
photographic mosaic of the base of Mount Sharp
highlights the layered nature of the lower mountain
and the differences in color and shape of the rocks.
ChemCam works primarily to identify chemical elements with its laser and spectrometer. But it also
has a telescopic camera to put its laser results
into a larger context, and it can be used independently to look for future targets at any distance.

a former director of JPL. What became known as Murray Buttes is a place of an importance that is both arbitrary and grounded in science: It's the beginning of the landscape and geology clearly defined by the mountain. It will still be several miles from Murray Buttes to the prime exploration area, but the terrain starts to change, with more mesas, buttes, ridges, and outcrops of all kinds.

An early target will be what is called the washboard unit, a visible-from-orbit formation of rock with distinctive parallel wavy rises that gave rise to its name.

The washboard bears some resemblance visually and in texture to a larger unit of rock near the Curiosity landing site. Team geologists have reason to believe the two may represent separate sections of the same rock layer, with the black dunes covering the rock that connects them. If that continuity can be established, then the process of connecting the Yellowknife Bay area with the mountain will get a huge boost.

After navigating through some troughs, the rover will come to one of the more remarkable features in the area: a 32-foot-high, 650-foot-wide, and 4-mile-long (10-meter-high, 200-meter-wide, and 6.5-kilometer-long) wall filled with the mineral hematite. It's a mineral often formed in water and is usually black, although much of this ridge has turned red.

Hematite has been identified before on Mars, but in a very different form. The Opportunity rover at Meridiani Planum found small balls of the mineral, which were nicknamed "blueberries." Like those hematite berries, Mount Sharp's imposing wall of hematite—a mineral that on Earth is mined for its iron ore—is especially strong and therefore resistant to erosion.

Hematite is not a mineral generally associated with habitability and life, but on Earth some of its precursor mineral states have an unusual relation with a particular type of microbe: the family of chemolithotrophs.

These microorganisms find their energy (food) in the form of electrons being gained or lost from the rocks around them. And in a 2013 paper about the Sharp hematite ridge, authors Abigail Fraeman and Ray Arvidson of Washington University write that these chemolithotrophs on Earth play a central role in the modification of iron, the process that leads to the formation of hematite. So the natural question that will be explored is whether the red ridge on Mount Sharp was once home to this kind of primitive organism, too.

The path up Mount Sharp will generally be a gradual one, in part because that's what the rover can handle and in part because much of the mountain rises at only a 10-to-15-degree incline. (Grotzinger likens it to Mount Fuji in Japan.) But the rover will sometimes follow an ancient ravine and find itself surrounded by steep cliffs. Some craggy and awkwardly placed outcrops of rock will also be tempting to explore, as will be pushing the envelope to see how much incline the rover can take while performing a drill. The steeper parts will be more hazardous, but they will also most likely be where the most exciting layering will be found.

Like many on the Curiosity team, Joy Crisp (left) and Ray Arvidson (right) are veterans of past NASA missions. Here Arvidson holds up a chunk of gray hematite to punctuate a daily status briefing to reporters.

"This is where we'll spend a lot of time, because there's just so much to learn. I have to believe there will be terrific discoveries and surprises."

In addition to the formidable hematite wall, ridges of rock known to be filled with clay and sulfate minerals rise out of the soil as the climb proceeds. These ridges, first called "fences" because they initially appeared to block passage, are the gold mine of Sharp. They reach 500 to 650 feet (150 to 200 meters) up the mountain and are the heart of the guidebook Grotzinger and other Mars scientists have been talking about.

The current plan is to go straight up the sequence of minerals, stacked (Grotzinger calls it one single stack) in layers, rather than going side to side. By reading the rocks from oldest (lowest) to youngest (highest), the team will determine the basic structure of the mountain and can learn how it was transformed over the eons as the climate and atmosphere changed.

Grotzinger's often intent and excited voice grew increasingly so as he described what lay ahead: "After we nail one stack, we can always move over and do the same for another. This is where we'll spend a lot of time, because there's just so much to learn. I have to believe there will be terrific discoveries and surprises."

A TALE TOLD BY BOXWORK

At the higher reaches of the lower mound is one of the more intriguing formations identified so far, called the boxwork. It was first discovered via HiRISE, but it has now been studied extensively by Grotzinger and Caltech graduate student Kirsten Siebach.

The boxwork is in effect a kilometer-wide lattice, an exposed and heavily eroded maze that tells a surprising geological story. Almost 800 meters (half a mile) from the base of Sharp, it represents the remnants of a sandstone formation where cracks opened in the rocks and then salts, most likely containing sulfate minerals, entered the fractures and formed much harder cements. Then the parent rock eroded and left the stronger, filled-in fracture lines to cement.

Siebach describes the boxwork as a flat but wavy, underground formation that had to have arisen in the presence of water. In this case, it could not have been water that covered the rocks, because they would have quickly dissolved. Rather, this was evaporating water coming up from the ground below what became the boxwork, supplying moisture without undoing the rocks and cemented cracks.

"What it tells us is that water was present this high up the mound, and that it probably came in the form of snow or rain," she said. "It seeped into the ground and then would evaporate up when conditions were drier. We can see only the exposed area, but it may well continue across the mound."

Because the boxwork formation is so high up the mountain, Siebach doesn't expect the rover to reach it for some years. It is also several miles farther horizontally on the gently sloping

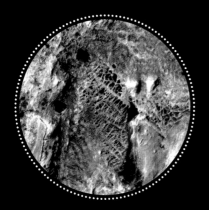 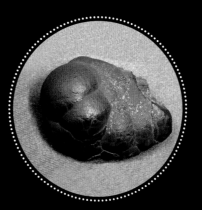 

MINERALS APLENTY >> Mount Sharp's geological treasure chest is known to have clay minerals in the lowest section of the mound, followed by sulfates, and then a break, with few minerals detected farther up. Hematite (above, center and right), the mineral form of iron oxide, is found near the base as well. Since the formation of these minerals happens in only certain conditions, reading the rocks tells Curiosity scientists a great deal about Mars history. At the flat area near the top, the unusual boxwork formation (above, left) has been identified by orbiting cameras. The model (below) was created from HiRISE images by the U.S. Geological Survey, with the height of the mountain exaggerated in relation to its width.

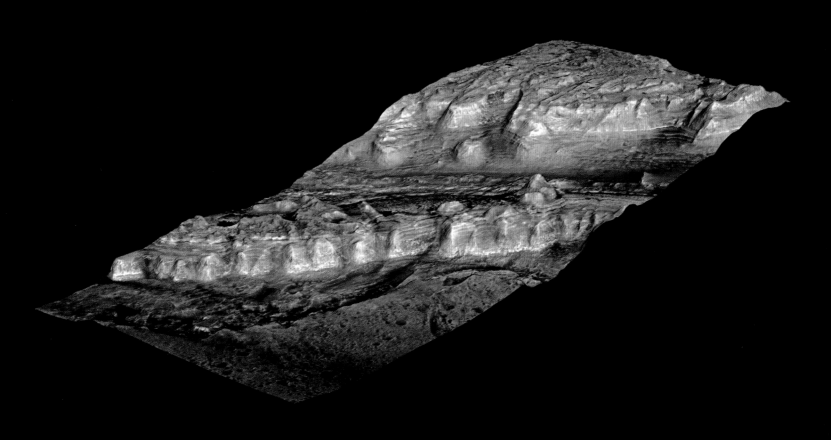

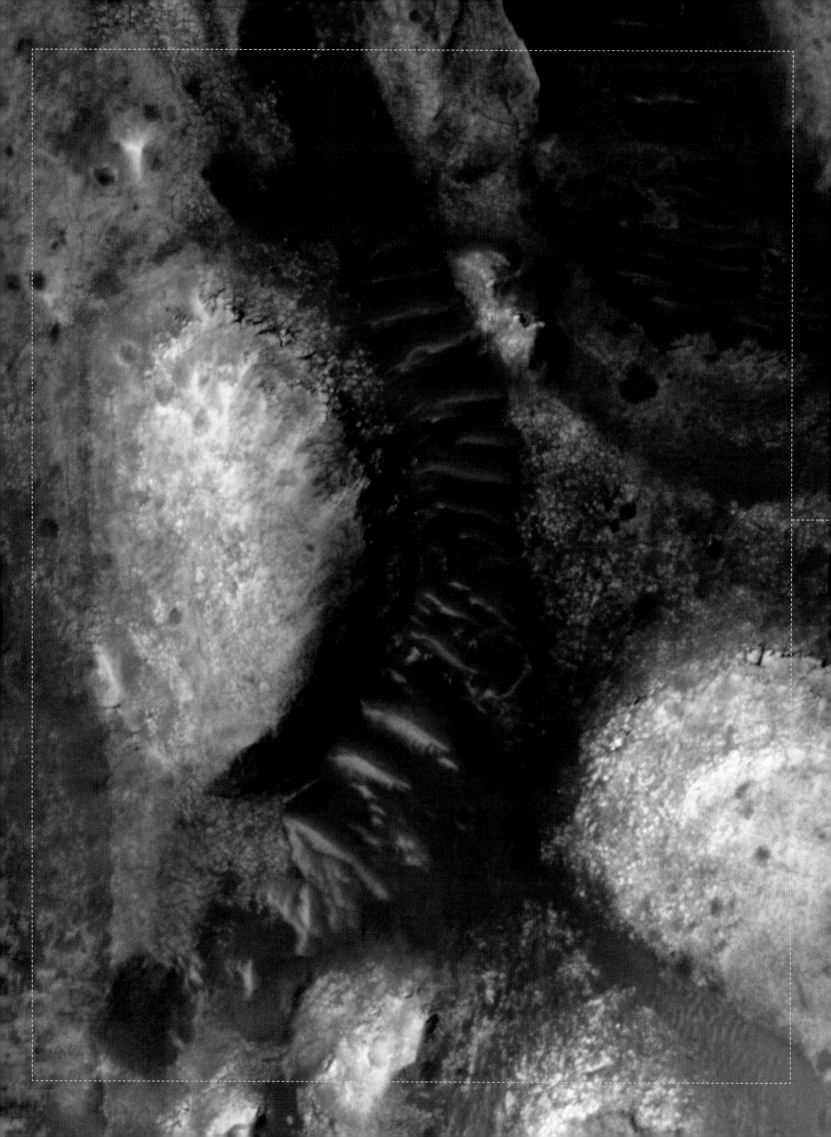

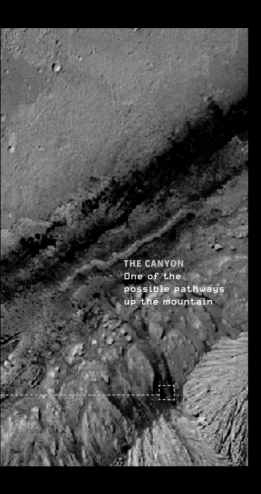

THE CANYON
One of the
possible pathways
up the mountain

A RICH ARRAY >> The base of Mount
Sharp is filled with science targets,
ranging from formed-in-water minerals to buttes, mesas, and deep
canyons. The rises on the left, ringed
by spits of sand in an ancient channel, are tiny details from a HiRISE
image. Looking down at the entry
point to the lower mound [above],
the large channel coming down off
the mountain is clear.

western side of the mound. But if the machinery continues to operate and the money to run Curiosity remains, then the plan is to approach and study the boxwork.

"It's on the list of targets to study, and there's a path up there," she said. "It would take a long time, but we *could* get there."

At almost one kilometer (about 3,000 feet) up the mountain, everything suddenly changes—a different kind of scalloped rock with faint or no mineral signatures takes over and continues rising to the top. This break, called an unconformity by geologists, is a likely place to stop—at least from a scientific perspective. The break itself would be interesting to study, but what's above is not considered to be of great scientific interest.

But that doesn't mean the rover won't climb farther. The nuclear-powered generator has enough plutonium-238 dioxide to keep the vehicle moving for more than ten years. So if other essential parts are still working after the science goals are met on the lower mound, then a decision could be made to push onward and upward, doing what science can be done and taking pictures the likes of which have never been seen.

AT THE CORE OF THE MISSION

Knowing that a site is a potential scientific gold mine is one thing. Figuring out how to explore it and dig out its secrets is quite another.

The rover planners were busy in the JPL Mars Yard with testing how to use the rover on much steeper inclines and more slippery surfaces, but the science team was also struggling with how to best continue its "journey of discovery" on Sharp. As Grotzinger put it, "We don't want to get there and just wander around."

So the project scientist put together a small team to help guide the decision making. It would propose priorities, identify promising sites, assess hazards, and come up with some possible pathways through the surrounding apron and up to the lower mound.

Grotzinger and Jennifer Trosper, by then the deputy mission manager, would lead the effort. Although Curiosity would arrive at Sharp many months after landing at Gale, Grotzinger says it nonetheless remained part of "our core mission."

"It's pretty clear that all eyes will be on us when we reach Sharp, and we don't want to leave any room for error."

"It's pretty clear that all eyes will be on us when we reach Sharp," he said, "and we don't want to leave any room for error. Basically, we want to come up with a selection of possible routes that the team can then weigh and ultimately select a path—pretty much like the way the landing site was chosen."

One of the team members is Bethany Ehlmann, a Caltech planetary sciences professor and research scientist at JPL. She has written extensively about the mineralogy of the Gale watershed and other Martian sites, and will be focusing on ways to analyze the ways the formed-in-water minerals are organized and the implications of that organization.

"There are seven key locations on Mars where there's an extensive and relatively exposed mineralogical sequence stacked on top of each other," she said. "We'll analyze the stratigraphy of Gale, and then figure out how that relates to the others. What is related? What conflicts? The answers will hopefully teach us which changes we see are local and which are global.

"We do the same on Earth. There's nothing here from the first billion years preserved like on Mars, but we can go to places like Western Australia, Greenland, South Africa, and northern Canada and see if the rock record shows big changes at around the same time. If it does, then you know something happened on a global scale."

Ehlmann will be joined by Milliken, who has also studied the Sharp mineralogy and layering for orbital information. Their job, Grotzinger added, would be to "look at every darn pixel we have to see which are the ones with the best chance for a clean signature." The signee here will be a deposit of clay or sulfate or another mineral formed with water, and the goal is to get the rover to the sites where the minerals are most clearly concentrated.

Grotzinger also put on the planning team a geomorphologist to guide thinking about the landforms, a geochemist, and a physicist. The team leader wanted people with different backgrounds, approaches, and understandings to "stir into a big pot of gumbo," to create a tasty and ever intermingling scientific concoction.

All the while, the primary goals remained the search for places where life might have once survived, or even flourished, and the elusive search for organic compounds. As had become increasingly apparent, the challenge was not only to find where they might have landed or been formed in place. Equally important was where they may have been preserved.

Also near the top of the science agenda was the issue proponents of Gale emphasized when they lobbied for it as the landing site: What was going on when the layers with clays stopped being formed and those with sulfates began? Sharp has places where the contact between the two layers appears to be close, and that proximity is a window into a potentially major turning point in Mars history.

Since clays and sulfates are formed in quite different conditions, something important had to happen to cause the change. Was it an abrupt transformation? If the line between the

>> AUGMENTED REALITY
View this image
through NASA's
Spacecraft 3D app.
Learn more on page 6.

"Big picture, we're trying to learn about
changing climates and conditions on Mars."

The history of Mars is written on rocks, and especially its history of water. Three to four billion years ago, that water was substantial enough to form large deposits of minerals that need water to assemble. Ehlmann and her colleagues analyze those ancient minerals from orbital data and by zapping Martian rocks with the ChemCam laser spectrometer aboard the rover, then reading the chemical "fingerprints" in the plasma created. Like so many of her colleagues, she has done fieldwork around the world. When one of the satellites orbiting Mars sent back data that suggested a particularly interesting dynamic, she and others took a backpack-size version of the instrument to Iceland to test out the process. In addition to serving as an assistant professor at Caltech and a research scientist at JPL, Ehlmann has been selected as a National Geographic Society "emerging explorer."

Bethany Ehlmann, California Institute of Technology

Because of her background studying signatures of minerals in and around Gale Crater, Ehlmann was chosen to help plan the science campaign for the rover's approach and ascent of Mount Sharp (below).

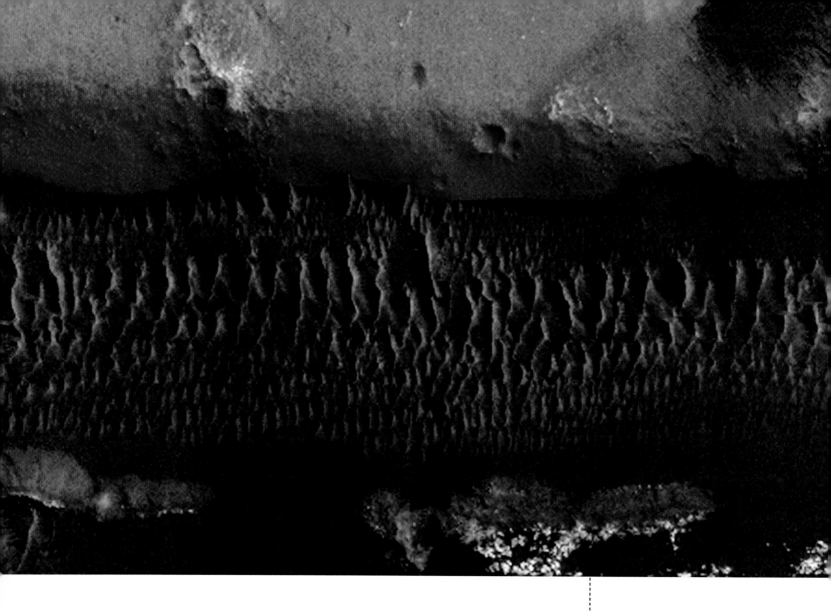

two mineral layers is clean and sharp, then it probably was abrupt. But if on examination it becomes more clear that there are sulfates in the clay layers and clays in the sulfate layers, then something more gradual happened.

And understanding the Sharp clays and mudstones would help the scientists put the clay and mudstones of Yellowknife into a broader context. Are they chemically and textually the same or similar to those on Mount Sharp? If the answer is yes, then they were more likely formed at the same time and are part of a continuous layer. If not, then they represent different periods when Gale was wet.

From some of these findings and measurements that perhaps seem less than dramatic, broad theories are born about the mountain, the crater, the planet.

The team would, of course, like to find a dinosaur bone or, far more realistically, signs that a colony of microbes once lived in a particular rock. But that wasn't going to happen, and the rover didn't really have the tools to make that kind of finding. So it was *CSI: Mars* again, putting together clues large and small that together could lead to a quite possibly revealing conclusion.

With a team of more than 350 scientists, there were naturally strong differences of opinion about what's important and worthy of Curiosity's time. With that in mind, Grotzinger ended an all-hands science team meeting in the fall of 2013 with a presentation and then a free-form discussion about how the team would attack Sharp.

Miles and miles of dark and sometimes shifting sand dunes skirt the base of Mount Sharp and form a barrier for the rover. But breaks in the dunes have been found, including a path of bedrock with buttes and mesas to the sides. Mars's black sands, in this photo at Lobo Vallis, are composed of volcanic basalt, as well as minerals such as olivine and pyroxene.

There were naturally strong differences of opinion
about what's important and worthy of Curiosity's time.

Opinions ranged across the board, with particular sites and approaches forcefully put forward. The clays and sulfates were the prime scientific target for sure, but the Curiosity mission had already shown that it was open to changes in plan. When pressed on this, Grotzinger gave this example: Pretend that on the way to the lower mound, the rover instruments detected the remnants of a hydrothermal vent—a break in the ground where superheated water once bubbled up or even spouted like the Yellowstone geyser. Vents like these are known to be hot spots for microbial life, and some scientists have theorized that life on Earth began at or near such a vent.

"Would we delay going to the minerals if we found a vent? You bet," Grotzinger told the crowd. "And if it was giving us great results, we'd stay for a long time."

The hydrothermal vent scenario is an unlikely one—they tend to be associated with volcanic regions—but the crowd got the point. The "mission of discovery" would continue on Sharp just as it had played out when the team decided on the Yellowknife detour. There was definite and growing pressure to reach and begin climbing the mountain, but competing pressures could also carry the day.

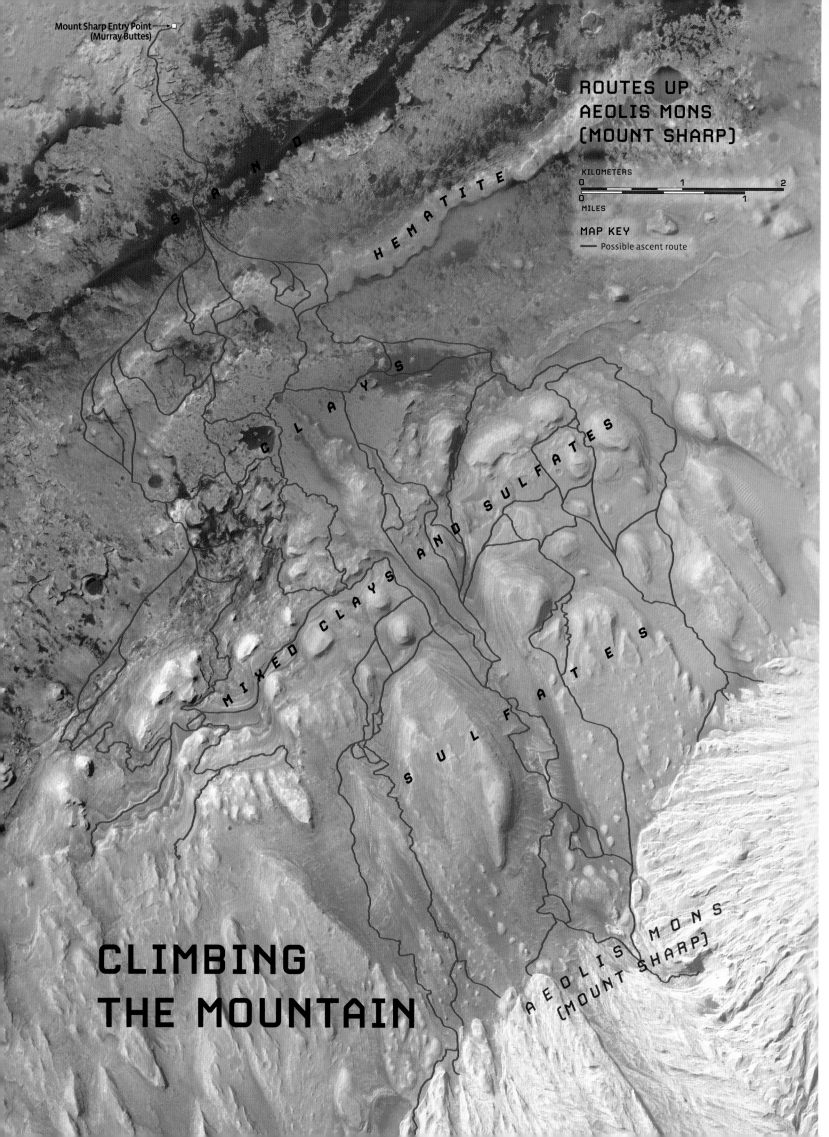

Mount Sharp Entry Point ──→
(Murray Buttes)

ROUTES UP
AEOLIS MONS
(MOUNT SHARP)

KILOMETERS
0   1   2
MILES
0   1

MAP KEY
—— Possible ascent route

S A N D

H E M A T I T E

C L A Y S

C L A Y S   A N D   S U L F A T E S

M I X E D   C L A Y S   A N D

S U L F A T E S

CLIMBING
THE MOUNTAIN

A E O L I S   M O N S
(MOUNT SHARP)

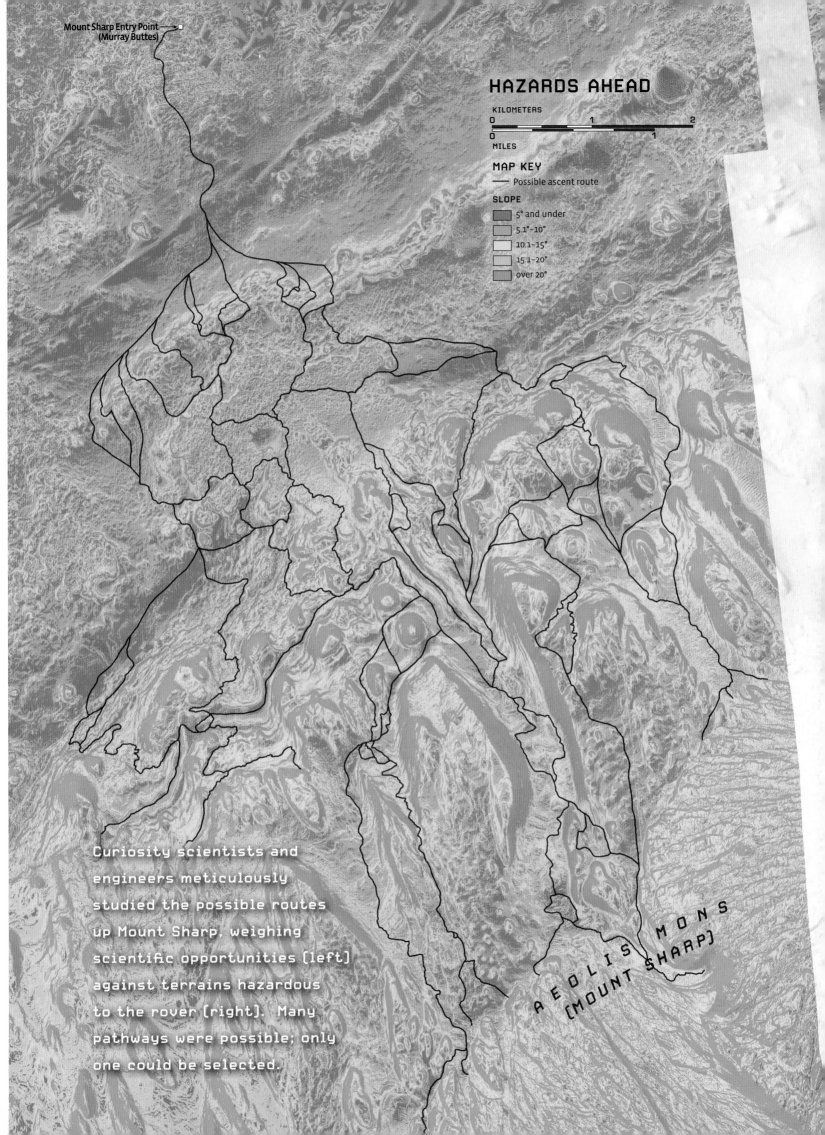

Mount Sharp Entry Point ⟶
(Murray Buttes)

HAZARDS AHEAD

KILOMETERS
0   1   2

MILES
0   1

MAP KEY
— Possible ascent route

SLOPE
5° and under
5.1°–10°
10.1–15°
15.1–20°
over 20°

Curiosity scientists and
engineers meticulously
studied the possible routes
up Mount Sharp, weighing
scientific opportunities [left]
against terrains hazardous
to the rover [right]. Many
pathways were possible; only
one could be selected.

A E O L I S   M O N S
[MOUNT SHARP]

FOURTH-PLANET
ASTROBIOLOGY

CHAPTER 12

If Mars once had much more water, might it have supported life?

BACK IN THE RUN-UP to the twin Viking landings on Mars, the ever enthusiastic Carl Sagan pushed hard for a mission that would look not only for microbial life but also for life you could see and, if on the planet, reach out and touch. The soon-to-be star of the television series *Cosmos* and the nation's leading voice on the search for life beyond Earth, Sagan was quite convinced that Mars did harbor evolved life and was impatient with those who were not.

His partner at *Cosmos* and previously on the Viking missions was Gentry Lee, a top JPL engineer, operations director, and later science fiction author. Lee is still at JPL, where he is a chief engineer for the entire Solar System Exploration Directorate and a prominent minister without portfolio who remains involved not only with engineering but also with promoting space travel and reviewing operations in progress (including Curiosity).

He's also a great, demonstrative, desk-pounding storyteller and a repository of tales of the early NASA. One of his favorite topics is Sagan.

As Lee explained it, Sagan was innovative and aggressive in promoting his scientific agenda. One of his tasks on the Viking mission was to determine how to look for life of various sorts on the Martian surface, and then to be part of the team doing the looking.

Not long before launch, Sagan learned that one of the onboard instruments that he had pushed for had been

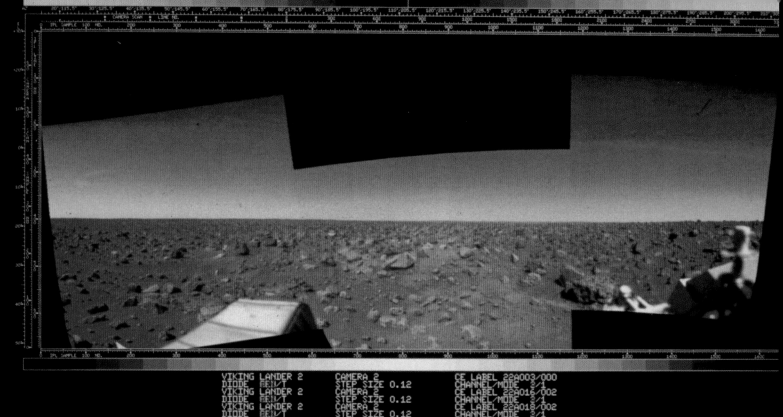

VIKING LANDER 2   CAMERA 2   CE LABEL 22A003/000
DIODE BB1/T   STEP SIZE 0.12   CHANNEL/MODE 2/1
VIKING LANDER 2   CAMERA 2   CE LABEL 22A016/002
DIODE BB1/T   STEP SIZE 0.12   CHANNEL/MODE 2/1
VIKING LANDER 2   CAMERA 2   CE LABEL 22A018/002
DIODE BB1/T   STEP SIZE 0.12   CHANNEL/MODE 2/1
COLOR MOSAIC OF RADCAM OUTPUT SPEC MIN 0. MAX 4.5 *
LABCAT
SAR - LGEOM
MASKVL
                        SEGMENT 1 OF 1
                IPL PIC ID 76/09/14/125932   WDB/L1473BX
                JPL IMAGE PROCESSING LABORATORY

One might imagine that the search for life beyond Earth might be seen as essentially quixotic. Not so.

FROM THE PHOTO ALBUM Images sent from Mars by Viking landers in the 1970s show a rocky plain and what appeared to be snow or ice on the ground (left).

scrubbed. It was a lamp that would have lit the night around the landers and allowed photos to be taken. But the $150,000 cost was ultimately deemed too high, or perhaps the idea was considered far-fetched. Lee took the story from there:

"When Carl learned about this, he stormed into the mission manager's office and started giving him hell. This was outrageous, a real blow to the mission. The managers should be ashamed. And then he said, and I remember this well, 'How will you explain the tracks left behind at night from the natural animals that were there?'"

FIRST GLIMPSE >> The first color image from the surface of another planet (opposite) came to Earth from Viking 1 in 1976. Scientists such as Carl Sagan, astronomer and central advocate for the mission (above, with Viking model), were optimistic that signs of life—even visible creatures—would be found, but instead the planet appeared parched, frigid, and desolate. One set of experiments measured what principal investigator Gilbert Levin remains convinced were indications of life. But other instruments came up with different results, and the confusion added to disappointment in the search for life on Mars.

Sagan was then the world's most prominent astrobiologist (or what was called an exobiologist at the time), and according to Lee he was deeply disappointed by the Viking results. His certainty about evolved life on Mars was undeniably misplaced, but it turned out that his interest in and optimism about Martian and other extraterrestrial life was shared by many and had a future.

Some of that interest has been fanciful for sure. As imagined, Martian life has ranged from the intelligent and telepathic peoples of writer Ray Bradbury to the evil creatures of H. G. Wells's *The War of the Worlds;* from the great canal builders envisioned by Italian astronomer Giovanni Schiaparelli and American Percival Lowell in the 19th and 20th centuries to the Mars "face" sculptors said to have chiseled rock creations for visitors to see from above; from a rodent supposedly photographed on Mars by Curiosity's cameras to Sagan's night-visiting creatures.

But finding signs of life beyond Earth has also been a goal and a hope for many scientists and many simply fascinated with space and what the discovery of life elsewhere might mean to us all. On Mars, this considered scientific search for life has focused in recent years on subjects far more modest than what Sagan envisioned—single-cell microbes that more likely existed billions of years ago than are still eking out a living today.

In the mid-1990s, for instance, NASA announced that its scientists had found signs of long-ago life in a Martian meteorite that had been found in Antarctica. A decade later, other NASA

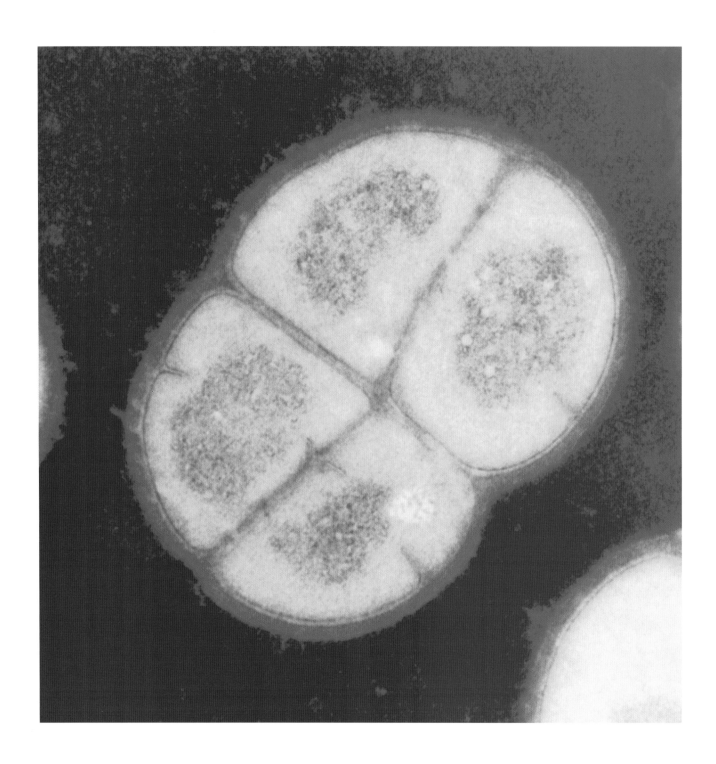

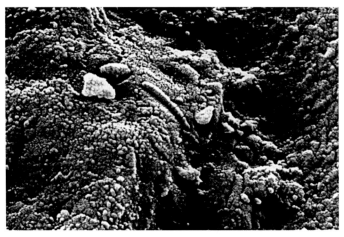

The extremophile bacterium *Deinococcus radiodurans* can withstand vastly more radiation than virtually any other known life-form, and as a result is of great interest to Mars scientists and astrobiologists in general.

<< The debate over the Mars meteorite ALH84001 focused most intensely on the claim by researchers (now largely rejected) that they may have found a microfossil in the tubelike structure photographed here with an electron microscope.

scientists announced findings that suggested primitive organisms might live deep under the Martian surface.

But the microbe findings have run into trouble as well. The Mars scientific community found fault with the Mars meteorite research, and the critics have largely won the debate. The most recent setback—or perceived setback—in the search for extraterrestrial life came at the hands of Curiosity when it was unable to detect the gas methane in the Martian atmosphere.

Testing for methane was a priority for the SAM team, and the instrument has sniffed the air periodically for the gas with little or no success. The reason for the effort was that several scientists, most prominently Michael Mumma of Goddard, had published papers describing episodic plumes of methane belching out from Mars or detectable levels of the gas in the atmosphere. On Earth, most methane is the by-product of living organisms, and inevitably some scientists put forward theories that the Martian methane was a signature of Martian life.

Mumma never made that direct connection, although he presented it as a possibility along with the possibility that the gas was produced via geological processes. But when SAM measurements, conducted by Christopher Webster of JPL, came up negative, the general story line was that yet another theory regarding life on Mars had been knocked down. Mumma stood by his findings that methane episodically was released in substantial amounts from the surface of Mars, but Grotzinger and the SAM team made clear that trying to find methane on Mars had fallen as a priority. It will be, however, for upcoming European and Indian missions.

ASTROBIOLOGY TODAY

Given this history of dashed hopes, one might imagine that the search for life beyond Earth might be seen as essentially quixotic. Not so. The field of astrobiology is actually booming and has an ever increasing plausibility to it. What's more, this coming-of-age has been based on major discoveries that are not controversial at all—breakthroughs that have come from the far corners of the Earth to the far corners of the universe.

On Earth, scientists have shown that single-cell life can live in the most extreme environments imaginable: in rock fissures three miles underground in South African gold mines, under Antarctic glaciers, when exposed to high doses of usually deadly radiation, even floating in the atmosphere. These discoveries in the world of microbes known as extremophiles have greatly expanded our idea of what a habitable environment can be.

In addition, astronomers and astrochemists have learned that all the ingredients for life are present throughout the galaxies. This chemical soup includes not just elements like oxygen, nitrogen, and carbon, but complex molecules of amino acids and nucleobases that are key to the workings of life's genetic code. These substances rain down on planets, moons, and asteroids in forms from meteorites to interplanetary dust particles.

And perhaps most persuasive, astronomers have identified thousands of planets outside our solar system and now are convinced that billions more exist. What's more, a considerable number orbit their suns in what is called the habitable zone, the distance from a sun in which the presumably essential solvent water will not always be frozen and will not boil away.

"Mars probably had the right conditions for an origin of life, but then it went on a very different path than Earth."

To those following discoveries like these, the implausible position now is that life exists (or has existed) only on Earth.

So while Sagan was dead wrong about creatures on Mars, he was on target when he popularized the admonition that "absence of evidence is not evidence of absence." It's a phrase often heard when scientists are struggling with the difficult process of learning what is and what is not the case far from Earth.

If an experiment does not come up with a result that, based on other findings, a scientist had hypothesized would be produced, does that mean the hypothesis is incorrect? Or does it mean that the instrument, techniques, and understandings needed for conducting the experiment were inadequate or inappropriate? Clearly, either could be the case.

A good example involves those planets that orbit suns other than our own, known as exoplanets. Astronomers theorized about their presence for centuries. But they were unable to produce evidence of their existence until the mid-1990s, when new techniques for detection were invented. Now finding another exoplanet is no big deal.

Similarly, many have theorized about the possibility that life once existed on Mars, or perhaps still does deep underground. Scientists have no evidence at this point to confirm that theory, but the Curiosity mission is nothing if not an effort to improve our ability to address the question with new tools, new techniques, and new understandings.

IF NOT LIFE, WHAT?

Officially, Curiosity is not a life-detection mission—it is not equipped to determine if Martian life exists now or ever existed. But it is an astrobiology mission, and NASA fully anticipates that it will lead, sooner or later, to missions that will be actively looking for signatures of Martian life present or past.

That focus on astrobiology is sometimes easy to miss. The team, for instance, was broken down into science theme groups that had been created before landing, and they ranged from geology and mineralogy to topography, from environmental and atmospheric science to chemistry—but not astrobiology.

Nonetheless, astrobiology has always been at the core of the Curiosity science mission. Jennifer Eigenbrode, the biogeochemist from Goddard working to find organics, described it this way: There wasn't any need for an astrobiology team because the entire Curiosity mission is about astrobiology. Finding habitable environments and organics are core science objectives.

 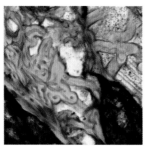

FOSSILIZED BACTERIA Microfossils from Antarctica and ancient lava flows are the kinds of evidence of life that researchers think could be on Mars.

"Mars probably had the right conditions for an origin of life," Eigenbrode explained, "but then it went on a very different path than Earth. So we can assume that if there was life, it would have evolved to adapt for however long it could. And we know now that life adapts to every extreme environment once it starts.

"Our job as astrobiologists on a Mars mission is to try to develop that story, pull the pieces of the puzzle together."

Eigenbrode paused and gave a sly smile. "Even cooler, what if life never happened there? Why not? It should be. Whichever path it is, it's a mind-shaking conclusion that we go to."

Some team scientists and some rover instruments are certainly less directly involved in what might be called astrobiology—those measuring wind and weather, for instance, or studying the geology that created Mount Sharp. But they are nonetheless part of a conscious strategy to broaden how NASA looks for evidence of past or present extraterrestrial life.

By measuring and analyzing more and different aspects of the Gale Crater environment and Mars generally, they are trying to be smarter and more realistic about the size of the challenge before them. The Viking missions famously came back with a muddied answer to the

Astronomers have concluded there are hundreds of billions of planets outside our solar system, and billions of them exist in what are called habitable zones, where liquid water can persist, making them more like a "Super Earth," portrayed on the right. Planets orbiting too close to or too far from their stars will be more like the dry, toxic wasteland of a "Super Venus," shown on the left.

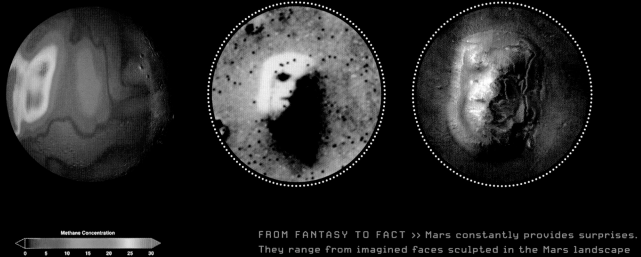

Methane Concentration

0 5 10 15 20 25 30

FROM FANTASY TO FACT >> Mars constantly provides surprises. They range from imagined faces sculpted in the Mars landscape (above, right, and center) to possible detections of the important gas methane (above, left) to strands of color that appear to be briny water flowing in summer down some crater walls (below). These mysterious seeping features were first reported in 2011 based on HiRISE images and have subsequently been found in hundreds of locations. Some scientists have suggested these flows could support Martian microbial life even today.

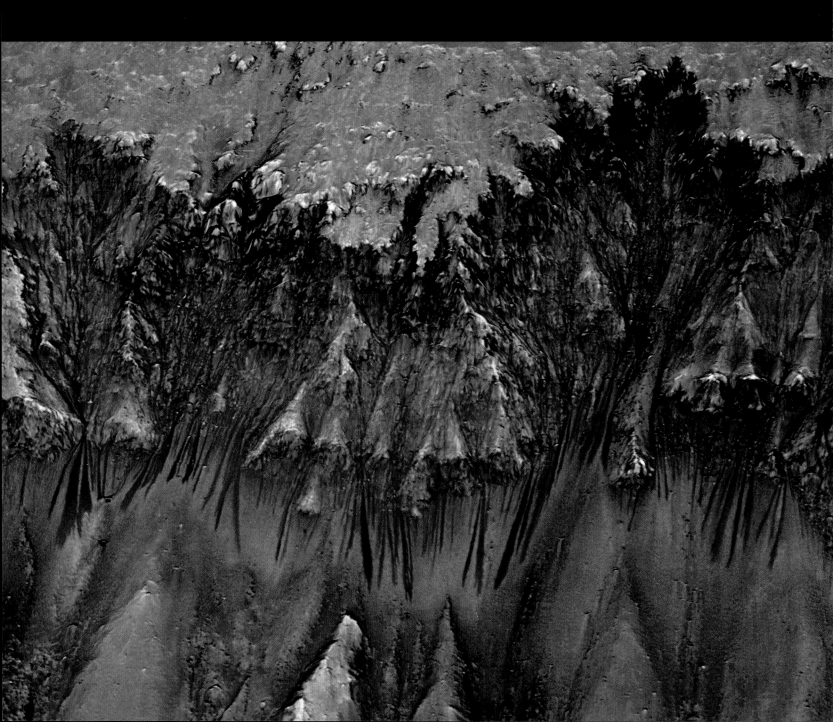

question of whether Mars harbored life; one experiment found evidence of metabolic activity, while others found that even simple organic material was absent. The confusion, the planetary science community now generally agrees, was the result of a certain hubris and overreaching. NASA was trying to run on Mars before it had learned how to even crawl.

Since then, Mars missions have sent back enough information to at least begin to answer some questions about the prospects for Martian life. Curiosity is filling in some of those gaps, while opening others.

GAUGING RADIATION

The rover, for instance, has the first instrument to measure radiation on Mars, the Radiation Assessment Detector (RAD). High-energy particles from solar flares and distant exploding stars are omnipresent on the Martian surface, and have been since the planet's early thicker atmosphere disappeared. They're known killers, breaking down chemical bonds in living organisms and simple organic compounds. Given this radiation assault, could anything resembling life survive?

RAD found that radiation *would* be a serious problem for any organism within one meter of the surface. But radiation modelers had previously concluded that no organism could survive in the first two meters of the Martian surface, so the new readings suggested a less severe radiation environment than modeled. The European Space Agency's ExoMars mission—scheduled to land on Mars in 2018—will come with a drill that can dig down two meters to a level that RAD principal investigator Donald Hassler writes "is potentially habitable with respect to incoming radiation."

Curiosity's Radiation Assessment Detector (RAD) instrument is collecting the first measurements ever on the cosmic galactic rays and solar particles that bombard the Martian surface. While radiation readings have been high, researchers say organisms like *D. radiodurans* could survive them.

What's more, some microbes on Earth are known to be able to survive high levels of radiation, and Martian microbes could similarly evolve to ward off even more. Hassler writes that the radioresistant extremophile *D. radiodurans,* for instance, could survive even in the top meters of Martian soil because it can go dormant for long periods. If it began life in the planet's early days—when the atmosphere was thicker and able to protect against radiation—it could potentially remain alive even today by awakening only in less hostile times and then repairing whatever radiation damage had been done.

Like RAD, the two rover laboratories—SAM and CheMin—are new and designed to fill gaps. It was primarily results from these instruments that led to the determination that Yellowknife Bay had once been habitable.

CheMin identified the formed-in-water minerals present and determined they came together in water that was neither too acidic nor too alkaline for life. SAM geochemistry measurements showed related elements and compounds at different states of charge, freeing up electrons that would be a source of energy for microbes. Add the finding that Yellowknife had been under water, and you have a place where microbes could have lived.

Astrobiology, the ultimate multidisciplinary field, takes it all in. If new information can shed any light on the question of whether life ever began on Mars or elsewhere, then it's

"Follow the water" has long been one of NASA's guiding principles for exploring Mars and searching for life.

astrobiology. The goal is to find extraterrestrial life, if it exists or once existed. But for now, the Curiosity team is mostly learning how to look.

"Follow the water" has long been one of NASA's guiding principles for exploring Mars and searching for signs of life. The ground-truthing of the Peace Vallis river and alluvial fan was a triumph of that approach, and the Curiosity mission is in many ways an effort to dive into the new "follow the carbon" formula.

But the story of Martian water certainly didn't end at Peace Vallis. Rather, the discoveries there ratcheted matters up: Where was the flowing water coming from? Was it rain? Snowmelt? The result of some underground geological or hydrothermal event that warmed subsurface water and started it flowing? The heat from a large meteor impact?

And if Peace Vallis's faint fan seen from orbit was confirmed to have been carved by water, what about all those other sculpted-in-rock fans, deltas, and possible lakesides? Many have formed-in-water mineral signatures around them, and Mars scientists can no longer assume that the lack of a mineral signature means that it isn't there.

Yellowknife Bay, after all, did not show any signs of mineral deposits, yet clays and sulfates were pulled up by the drill. Dust had hidden those deposits from the orbiting satellites, and no doubt does the same elsewhere on the planet.

So it is no longer controversial to say that while the surface of Mars is very dry now and has been for some time, the mineral, rock, and landscape evidence tell of a time when the planet was much wetter. SAM analyses of water tell a similar story.

BACK TO THE L-WORD

A team led by Laurie Leshin, formerly a top NASA official and now dean of the school of science at Rensselaer Polytechnic Institute, found that Martian water locked into rocks and in the atmosphere is disproportionately heavy water. That means it has an extra neutron in its hydrogen component compared with lighter water, which is the norm.

If an abundance of heavy water remains on or near the ground, it means a lot of lighter water evaporated away. On Earth, evaporation is part of the water cycle, and the water comes back down as rain. But on Mars, the water began to float off from the planet once the atmosphere got too thin to hold it.

Mars water his at the center of efforts to understand Gale Crater as well. Some of the science—and emotion—associated with it broke out at an all-hands Curiosity science team meeting at Caltech.

It was pretty late in the day, but the large Baxter Lecture Hall remained filled with scientists. The lights had been dimmed, and onstage Bill Dietrich was telling several hundred of his colleagues

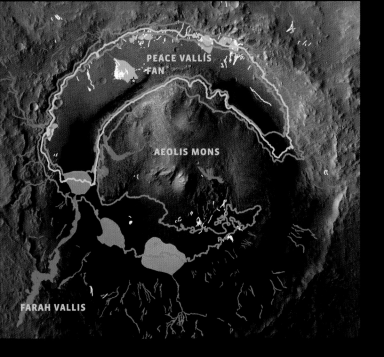

PEACE VALLIS
FAN

AEOLIS MONS

FARAH VALLIS

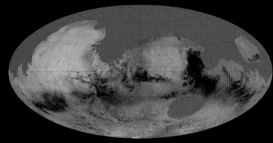

MAPPING ANCIENT WATERS >> For more
than two decades, Mars scientists have
debated whether a large ocean once
existed in the lowlands of the northern
hemisphere (hypothetical map, above).
Combining topographic and photographic
images of Gale (left and below), researchers examine the areas where streams and a
lake may have existed, and where features
suggest a past with deep waters.

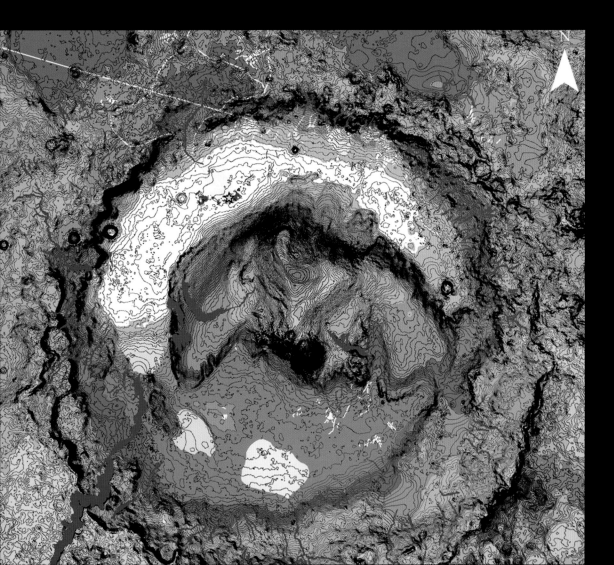

N

"We're reaching to make the greatest discovery of any kind: That life, or evidence of past life, was found somewhere other than Earth."

Lee was director of Science Analysis and Mission Planning for the Viking landers in the mid-1970s and chief engineer for the later Juno mission to Jupiter, and he's still at JPL and has been a reviewer of and advocate for Curiosity. In addition to his work on missions, Lee is a longtime writer of science fiction, having co-authored the Rama series with Arthur C. Clarke. He was also Carl Sagan's partner in creating and designing the documentary series *Cosmos*. As a senior JPL figure without duties on specific missions, he is brought in to assess progress and says he's free to say whatever he wants—which he does.

Gentry Lee, Jet Propulsion Laboratory, NASA

Chief engineer for the Planetary Flight Systems Directorate at JPL and a science fiction writer, Lee looks at the mission from a "big picture" perspective—and asks the question of how much early Mars resembled early Earth (artist's rendering below).

that it was time to seriously address the possibility that the 95-mile-wide crater, which had rivers and streams coming into it, was home to not only a shallow lake but, at some point, a deep lake as well.

To support his view that a deep lake likely did exist, he showed HiRISE and other satellite images of the sculpted remains of what he identified as numerous river channels and fan systems ringing the crater. That interpretation was not controversial.

But then he also pointed to what he described as drop-offs that fell into formations that he identified as delta fans. Unlike an alluvial fan like Peace Vallis's, which was created by the flattening out of a stream or river, the characteristic shape of a delta fan is formed when a waterway empties into standing water. The size and contours of some of the Gale delta fans—especially one called "the pancake"—suggested to Dietrich and his colleagues that a lot of water had been once pouring into a pretty deep lake.

He also produced detailed and colorful topographic maps, drawn largely by colleague Marisa Palucis, that showed the outlines of a sizable basin at the foot of Mount Sharp. Channels appeared to run off the mountain as well, sometimes ending substantially above the crater floor and consequently suggesting a deep lake.

"I think we need to address the lake issue head-on," Dietrich told the science team. "The topography is telling us there was a substantial basin here, and that may well be an important part of our story."

That story already did include a lake of some limited size and depth at Yellowknife Bay, but what Dietrich was describing was much larger. There was some palpable resistance in the audience, but what he proposed was not only based on the mapping and analysis of the landscape he presented. A deeper lake would also help explain the concentrations of formed-in-water minerals going up Mount Sharp, as well as the boxwork formation on the mountain studied by Kirsten Siebach and Grotzinger.

But a deep Gale lake has scientific implications, not only in terms of water on Mars but also how Mount Sharp might have been formed. Scientists can look at the same images and information and come up with very different conclusions. This time it was Ken Edgett, the geologist and orbital camera specialist who has studied the surface of Mars intensely for the past two decades, who saw things differently.

His voice rising, he objected to Dietrich's analysis, or at least parts of it. It wasn't that he didn't agree that Gale may well have once had a substantial lake; it was more that his experience with Mars had taught him to always be cautious and always look for alternate explanations. What you think you're seeing may be quite different from reality. Eons or erosion and other landscape-changing forces can make an outcrop look like the mouth of an ancient delta when in fact it's merely a run-of-the-mill outcrop.

What's more, he had spent years studying the peculiar landscape of Gale Crater and other craters nearby, and knew that whatever was happening wasn't simple. Some of those other craters in the region were fully buried with sediment, some were half buried, some had channels going both in and out of the crater, and some had fossil fans that themselves seemed to be partially covered and partially excavated by the wind and other forces. So you can't understand Gale, Edgett says, without looking at and trying to understand those other craters, too.

THE NORTHERN OCEAN HYPOTHESIS

His views were tempered as well by the great northern ocean debate.

The hypothesis that much of the northern hemisphere of Mars was once an ocean has been a source of sometimes intense disagreement since the late 1980s. The idea originated in the observation that bedrock in much of the northern hemisphere is 1 to 3 kilometers (0.6 to 1.9 miles) lower than in the southern hemisphere; on Earth, such a large basin would be filled with water.

It's a dramatic and defining interpretation, and it has gained more or less credibility with new observations and discoveries. The original papers included descriptions of what was interpreted to be sections of ancient shoreline, but later and better images were inconsistent with that interpretation. As Edgett later explained to Dietrich and others, it was this kind of controversy that he had in mind when he challenged the notion of a large Lake Gale.

But the northern ocean debate is far from over. In 2013, a large fossil valley network and delta system was identified from orbital images by Roman DiBiase and a team from Caltech. The system showed many inverted channels draining into the delta of the system, which appeared to have a sharp drop-off. That kind of feature, DiBiase said, is associated with a river emptying into a large body of water, and in this case it emptied into what is now the northern plains but was perhaps once that northern ocean.

While the northern ocean is not a focus of the mission, Curiosity appears to be adding to, rather than subtracting from, its plausibility.

The support comes from the dynamics of Peace Vallis itself. As described by Dietrich, Becky Williams, and others, the waterway needed considerable liquid to produce the rounded rocks, the load of sediment, the conglomerates, the inverted channel, and the broad fan that they found. And the water likely flowed for tens of thousands of years.

To constantly recharge it and potentially many other rivers and streams, that region of Mars needed a water cycle to produce rain or vast amounts of snow to melt. And the way to make that happen is a very large body of water.

"Basically, I would say that to have a Peace Vallis flowing for a long time, you need a northern ocean," said Michael Mischna, an atmospheric scientist working with the Rover Environmental Monitoring Station (REMS) on Curiosity. "It's a necessary condition because you need that much moisture to produce the precipitation you need."

If the water source was local and limited, his models show that over time it would all migrate from the warmer equatorial regions to the poles. A large body of water like an ocean would be warmer and could logically source a continuing water system, with rain or snow, evaporation, collection into clouds, and then more rain or snow.

Mischna was not saying that a northern ocean necessarily existed—his expertise is atmospheres rather than landforms and the water on them. Rather, his point was that if Mars once had many substantial valley and crater water systems, it needed a large and stable source of standing water to feed them.

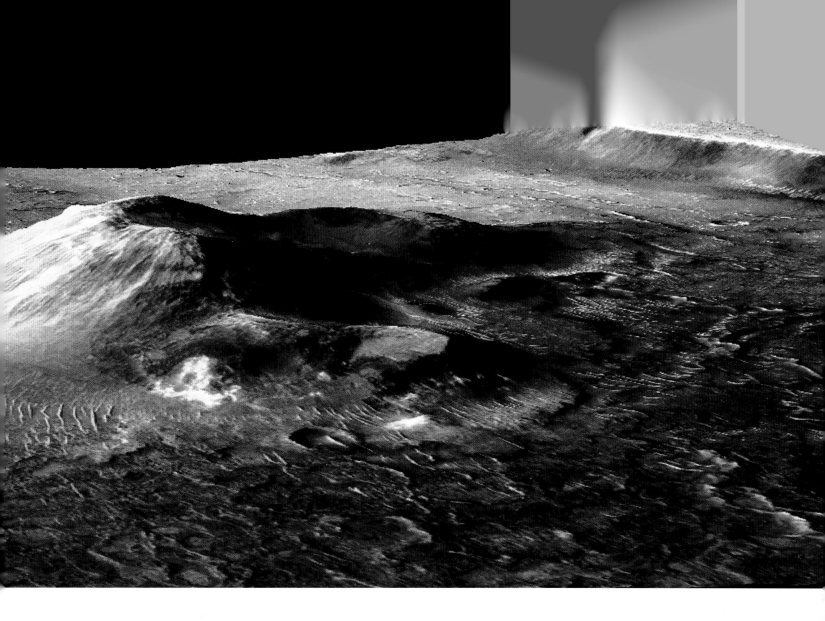

Some origins-of-life
scientists contend that life
began on Earth in steamy
hydrothermal environments—
in oceans and around
volcanoes. The volcanic cone
of the Nili Patera caldera on
Mars has signatures of
a similar hydrothermal past.

>> Astrobiology researcher
Lynn Rothschild studies
bacteria alive on Earth that
could withstand extreme
environments, including deep
space. Microbes that can
survive drought, extreme
cold, and high salinity are
considered candidates for
deep space survival.

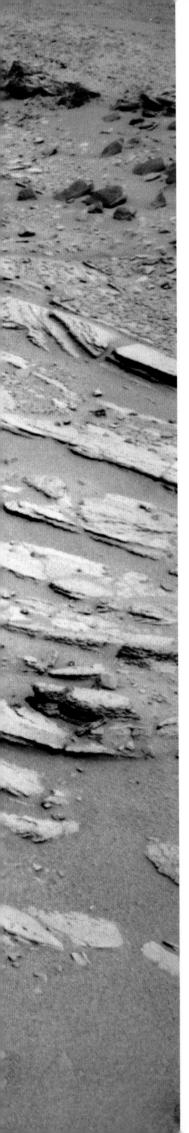

<< Layered sheets of rock hint at a watery past, suggesting Yellowknife
Bay was once habitable. This image captures both the terrain and the
rover itself, as well as tracks left behind by Curiosity.

For climate modelers, this remains an enormous obstacle because they still cannot produce a Martian atmosphere warm and thick enough to allow for surface water beyond the planet's earliest days. Clearly, this makes the existence of a northern ocean especially hard to imagine. But unlike some modelers, Mischna's view is that the evidence for Martian streams, rivers, deltas, and fans is strong and getting stronger. His discipline, he says, needs to work to explain how that could be, rather than argue why it could not be.

As the story of Mars water gets pieced together, it has become increasingly clear that the planet had a wetter history than imagined post-Viking. The early Noachian era had enough water to form many river valleys. Much later, what are called "catastrophic floods" opened vast canyons after lake water or melted groundwater broke through natural dams and cut the land. Now evidence for flowing and standing water during additional periods was impossible to ignore. This conclusion has potentially great implications for astrobiology.

Water is considered the optimal and perhaps essential solvent for life, and Mars once had a lot of it. At Yellowknife, the Curiosity team concluded that the planet—or at least parts of the planet—had all the other needed ingredients for life as well.

As the projected ancient Mars water budget rises, it becomes easier to entertain this possibility: For long periods during the early history of Mars, the planet was enough like early Earth that both could have supported an origin of life. The planets are very different now, but that doesn't mean they were very different four billion years ago.

Since the question of how life began on Earth remains unanswered, this is a comparison with a lot of variables. But still, it resonates with some—JPL chief engineer for solar system exploration Gentry Lee, for instance.

He says that the Curiosity discoveries already, and especially at Yellowknife, are a "very big deal" that need to be conveyed more clearly to people. This is what he wants mission leaders to say: "The public needs to know not only that they've found a place on Mars that was habitable. They need to know that it had a particular mixture of chemicals and particular minerals. And that if they ever appeared in that kind of juxtaposition on Earth, then there would be life.

"Nobody on the mission is going to say this because they want to manage expectations. I understand that. But as a person involved his whole life in the exploration of Mars, I can say it. And I can say we're reaching to make the greatest discovery of any kind—that life, or evidence of past life, was found somewhere other than Earth.

"If and when that happens, the man on the street will understand that something important has changed. He will look up at the night sky and what will he see? He'll see life. He'll look at all those stars and wonder how many more have planets around them with life, too. Really, that's what this is all about."

ALLUVIAL FANS AND ANCIENT DELTAS

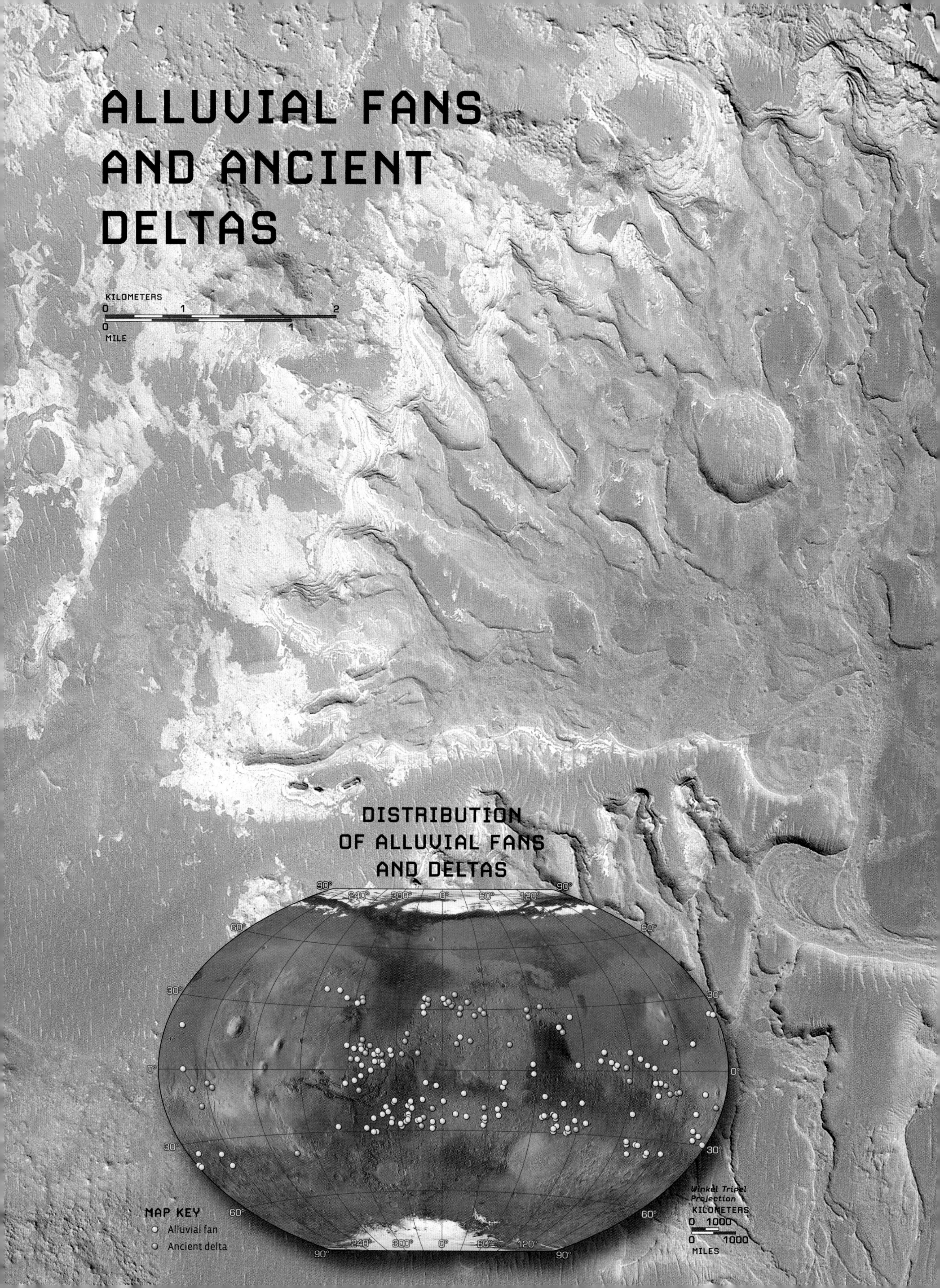

KILOMETERS
0
1
2
0
1
MILE
DISTRIBUTION OF ALLUVIAL FANS AND DELTAS
90°
240°
300°
0°
60°
120°
90°
60°
60°
30°
30°
0°
0°
30°
30°
60°
60°
90°
240°
300°
0°
60°
120°
90°
MAP KEY
Alluvial fan
Ancient delta
Winkel Tripel Projection
KILOMETERS
0
1000
0
1000
MILES

Eberswalde
Crater

This fossil delta in Eberswalde Crater testifies that Mars once had flowing water on its surface. It is not an isolated feature; the map at left shows other flow landforms that have been documented in craters across the planet.

HUMANS TO MARS

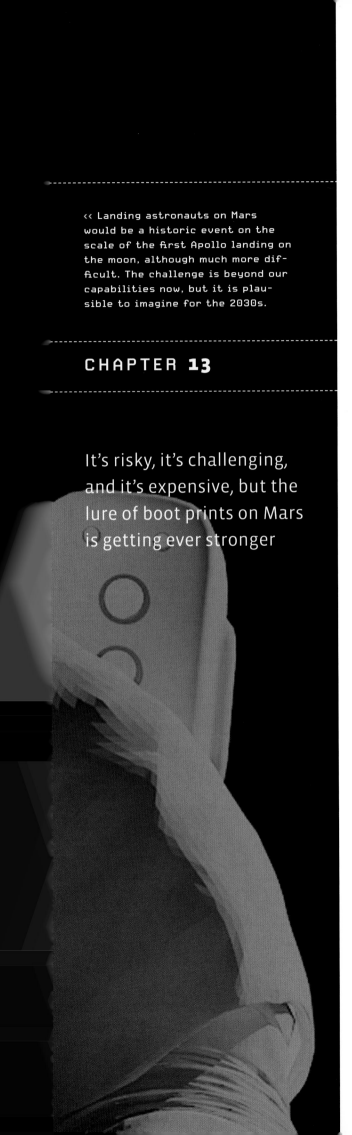

<< Landing astronauts on Mars would be a historic event on the scale of the first Apollo landing on the moon, although much more difficult. The challenge is beyond our capabilities now, but it is plausible to imagine for the 2030s.

CHAPTER **13**

It's risky, it's challenging, and it's expensive, but the lure of boot prints on Mars is getting ever stronger

H O W ' S   T H I S for a reality check?

The discovery of ancient Martian mudflats where life could have once existed is impressive. Ground-truthing the theorized presence of Martian rivers and alluvial fans in Gale Crater is impressive, especially since similar fossil rivers and streams around the planet are now more confidently presumed to have once carried water. Digging up clays and other minerals formed in the presence of water is impressive.

Untangling the geological and atmospheric history of Mars through the study of rock layers and units is impressive. Getting tantalizingly close to identifying organic compounds on the planet is impressive. Just driving a one-ton rover around Mars season after season is impressive.

They're very hard-won achievements and represent groundbreaking science and engineering. Yet as the Curiosity scientists and NASA officials know well, all but the last accomplishment represent the kind of achievements that a field geologist on Mars could claim in a small fraction of the time that Curiosity has been roving Mars.

The science and technology are cutting-edge, but really only because they're playing out on Mars. And as for climbing Mount Sharp, it will take the rover years to do that (if it survives), yet a human could potentially make it up in just a few days.

After three seasons on Mars, the head of NASA's science directorate, former astronaut John Grunsfeld, put it this way: "At this point we've driven about a kilometer, which an astronaut even in a space suit could travel in tens of minutes. They could stop and look and take samples like Curiosity does, and then bring them back to a lab at their base and get their results while eating dinner. I would say that in a couple weeks, they could do more science than Curiosity can do during its full one-year Mars mission."

And that pace of discovery is a significant improvement on the earlier rovers, Opportunity and Spirit. Humans, with their mobility, intuition, and ability to make connections and spot the unusual, have capabilities that robots don't and probably never will have.

Grunsfeld also recalled asking the principal investigator of the Mars Explorer Rover missions, Steven Squyres of Cornell University, how long it would take a field geologist on Mars to come up with the scientific results that Opportunity achieved in its first 90 days, which was its planned mission. Grunsfeld says Squyres stroked his chin, thought a while, and replied: "Twenty minutes."

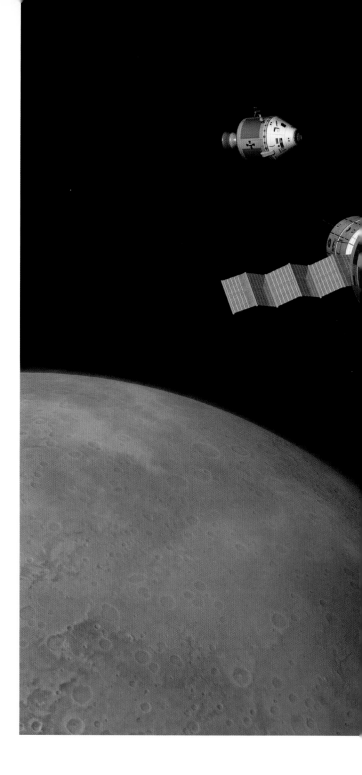

A foundation led by space tourist and investment tycoon Dennis Tito is working to send a man and woman round-trip to Mars in 2018. An artist's rendering shows a spaceship that would orbit the planet and return to Earth.

<< The BioSuit™ is a "second-skin" space suit that would allow for greater freedom of movement than many now in use. It was invented by MIT space engineering professor Dava Newman, who models one version.

Given these realities, NASA is eager to send humans to Mars and is actively planning a sequence of missions that could end with American boots on the Martian surface.

For the science, for the exploration, for the inspiration, that mission is a lodestar, and not just for the agency. Three private groups have also begun developing rockets and life-support systems to send humans to either orbit Mars or actually land on it. Some envision large colonies on Mars someday as humans become multiplanetary beings.

But as agency leaders often say, "Mars is hard," and it will be many years to decades before the journey can plausibly be accomplished. Difficult issues need to be resolved about development and construction of a rocket that is sufficiently powerful to lift a large, Mars-bound spacecraft off Earth; unanswered questions remain about how to land a large, peopled spacecraft safely on Mars and later lift them off; enormous challenges await those tasked with keeping astronauts healthy and alive both in transit and on the surface. Perhaps most sobering, it costs

Humans just can't always be in charge.
"The spacecraft will always be a day ahead,
and Curiosity has shown how to manage that."

many times more to prepare and execute a human Mars mission than to send a rover. The result is a pace of development that many Mars enthusiasts find frustrating in the extreme.

Nonetheless, the Curiosity mission is very consciously a stepping-stone on the way to sending humans to Mars.

That is most apparent in the inclusion of the Radiation Assessment Detector (RAD), the first instrument ever sent to Mars to directly measure radiation on the planet's surface. One of the early hurdles that must be cleared before NASA or others can seriously plan to send astronauts to Mars is fully understanding the hazardous, high-energy galactic cosmic rays and solar particles that bombard the planet and any traveling spaceship. Before Curiosity landed, it was unclear whether high radiation would be a deal breaker before there even was a deal.

Other Curiosity contributions to the human exploration program are less straightforward. For instance, in order to bring astronauts to Mars, the JPL Entry, Descent, and Landing team would have to be far more precise about where the capsule would land. The Curiosity landing ellipse, the designated area for touchdown, was 12 miles long by 3 miles wide. The successful landing was hailed as a major advance and achievement, since the previous rovers, Spirit and Opportunity, each had ellipses of 100 miles by 10 miles.

Progress for sure, but a landing ellipse for astronauts would have to be considerably smaller—maybe 1,000 square feet. This is because they would almost certainly be landing near

power sources, habitats, and facilities to make fuel, water, and oxygen out of Martian resources—all of which would have been preplaced in earlier robotic missions. Starting a human mission on Mars with a several-mile (or many-mile) trek to home base is considered highly undesirable.

The sky-crane landing was also an important innovation, made necessary because Curiosity was so much heavier than previous landers—one ton versus 374 pounds for Spirit and Opportunity, 900 pounds for Phoenix, 1,300 for each Viking lander. But for a human mission, the payload to land would be as much as 40 tons.

That means NASA and JPL will have to find new ways to descend and land. First would come heavy equipment, like machinery that can produce rocket fuel from components on the Martian surface. Then housing. Humans would come later.

WHY GO TO MARS?

Jason Crusan is director of NASA's Advanced Exploration Systems Division, and his list of Curiosity contributions to someday sending humans to Mars is long. But most important to him are the fault-protection and autonomous-activity advances made necessary by the time delays from Earth to Mars. As highlighted during the anomaly that almost killed Curiosity and hundreds of other occasions when the rover turned itself and its instruments off, humans just can't always be in charge. "The spacecraft will always be a day ahead, and Curiosity has shown how to manage that," he said.

Equally important to Crusan is the culture change at NASA regarding exploration and science. The two divisions have been historically separate in terms of planning and budgets, but now Crusan meets regularly with James Green, the agency's Mars program science lead, to coordinate activities and plans.

The 2020 follow-on mission of a Curiosity 2 takes the collaboration further: One of its primary goals is to test technologies that will someday allow astronauts to produce on Mars the water, breathable air, fuel, and food needed to live there.

There's been progress for sure, and a seemingly endless succession of enormous rocketry, engineering, and life-support problems lie ahead. But perhaps most daunting, there remains no clear commitment from government or the public to spend the potentially several hundred billion dollars needed to mount a humans-to-Mars campaign.

That's why the NASA humans-to-Mars program has been designed to play out step-by-step—accumulating knowledge from many intermediate missions. There won't be any Apollo-style crash program or timetable because there's no space race to Mars. Despite the growing spacefaring ambitions of China, India, and the Europeans, the United States has no realistic competition when it comes to landing on Mars now or in the foreseeable future.

Mars, however, has set an astronomical timetable of its own. Because of the shapes of Earth and Mars's orbits of the sun—one largely circular and the other more elliptical—the distance between the two can be great. At its shortest, the distance is some 35 million miles; at its greatest, it is 248 million miles. The average distance is about 150 million miles.

Those closest passes come at regular intervals—in cycles of about 16 to 18 years. The next period of proximity is 2016–2018, and that's why several Mars missions are planned for that period, most importantly the ambitious ExoMars landing planned by the European Space Agency and the Russian Space Agency. The next optimal time for a humans-to-Mars mission is in the early to mid-2030s, and that's why President Obama and NASA leaders talk so consistently about this time period as the logical time for a first attempt.

Even the 2030s are an overly long way off for Mars enthusiasts, but many are nonetheless heartened by what they see as a gradual change in thinking about the planet.

<< To learn about long-duration isolation, the Flashline Mars Arctic Research Station (sponsored by the Mars Society) in Canada's Arctic becomes home to scientists for months at a time.

Earth Launch
5 January 2018

228 days
outbound

Sun

Earth Return
21 May 2019

Venus when
Craft is at X

Venus' orbit

273 days
inbound

Earth's orbit

Mars' orbit

Mars fly-by
21 August 2018

Inspiration Mars aims
to send its flyby mis-
sion to Mars in early
2018, when Mars and
Earth are at their
closest. The 501-day
mission would use a
trajectory that, after
circling Mars, would
propel the spaceship
home without any need
for refueling.

>> Solar particles
from massive erup-
tions of solar winds
and magnetic fields
will be one of the
greatest hazards for
astronauts travel-
ing to Mars. While the
eruptions are infre-
quent, they send large
and damaging amounts
of radiation. In an
ultraviolet image such
as this, magnetically
active regions of the
sun burn brightly.

The issue now isn't so much whether humans will someday go to Mars, but when and how it will happen.

NASA is by all accounts the indispensable player when it comes to future human travel to Mars. The agency has the knowledge, the workforce, and potentially the access to the many tens of billions of dollars needed to mount a serious humans-to-Mars campaign. And sending missions to Mars is something about NASA that makes the agency unique: Seven vehicles have successfully landed on Mars, and all were sent by NASA.

Sending astronauts to Mars is also national policy. Speaking in Washington in 2013 at a conference organized by ExploreMars, a group that advocates for sending astronauts to Mars, NASA admin- istrator Charles Bolden made it clear. "A human mission to Mars is today the ultimate destination in our solar system for humanity, and it is a priority for NASA," he said. "Our entire exploration program is aligned to support this goal."

But it's no longer the case that the agency is the only player, or that the agency will neces- sarily even be the first to land an astronaut on the planet.

Other nations have Mars programs, but the most pressing competition comes from else- where—from wealthy private individuals, foundations, and commercial space companies. For love of exploration, challenge, perhaps glory, and maybe someday profit, a growing number of organizations have announced plans to attempt human missions to Mars as well.

PIONEERING SPIRITS

This is not space tourism they're talking about—like the Virgin Galactic flights that are sched- uled to soon take passengers briefly into weightless space, with some being serenaded by Lady Gaga as they untether. And this is not like the $20 million trip to the International Space Station that the first space tourist, financial adviser and billionaire Dennis Tito, took in 2001. These are plans for real Mars exploration, however plausible or implausible they may be.

Tito, it has become clear, is more than a space tourist; he wants to be a space pioneer. He is the driving force behind the Inspiration Mars Foundation, which announced in 2013 the goal of sending two astronauts—a man and a woman—on a Mars flyby in 2018. He has commit- ted to paying for the first two years of research and development, and initially he said he hoped to raise much more from private sources. Later his team said that for the plan to work, NASA would have to play a major financial and technical role in the flyby.

Like those of other private Mars ventures, Tito's goals are to increase interest, with the ultimate objective of setting up a colony on Mars. The technology is available, he says, if only the will (and resources) were available, too.

"I can't wait until 2030," Tito told *U.S. News and World Report*. "That's too long of a time to maintain enthusiasm. I think if we're going to fly to Mars, we have to do it with a short sprint to

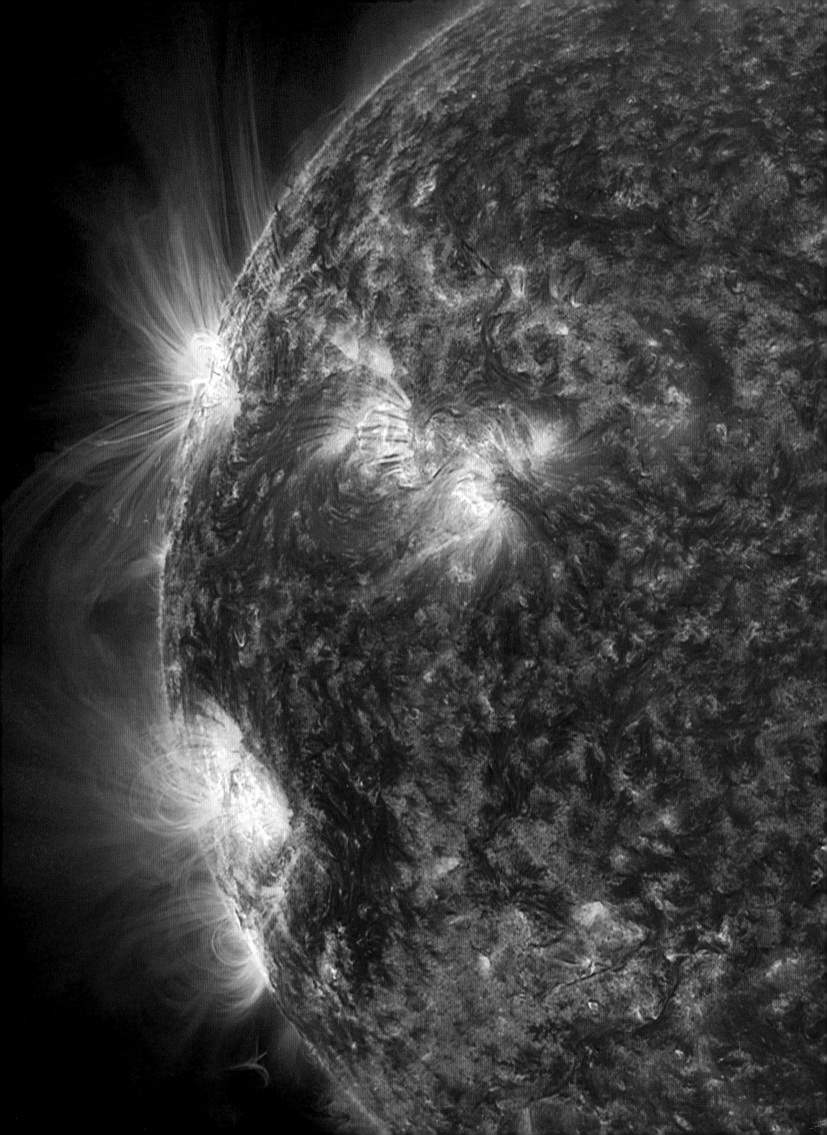

"We're the first instrument on the surface of another planet to measure radiation."

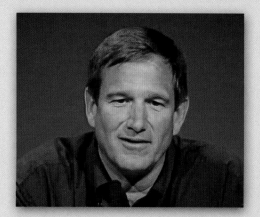

Donald Hassler, Southwest Research Institute

Hassler is principal investigator for Curiosity's Radiation Assessment Detector. We on Earth are protected from the radiation that the instrument measures by our magnetic field. The northern lights (below) are a manifestation of that protection.

Hassler has more than 25 years of experience in space physics and the development and calibration of space instrumentation. He is senior program director in the Planetary Science Directorate of SwRI, and he and his institute were selected to develop and build the RAD instrument in 2005. Hassler is also principal investigator for a second-generation RAD that will be sent to the International Space Station to measure long-duration radiation exposures there. The readings en route to and on Mars, as well as those made on the ISS, will be instrumental in determining the risks and possibilities of long-term stays beyond Earth. Hassler's involvement in radiation detection stems from an interest in space weather and the solar and planetary forces that drive it.

show we can do it and then we can take the time necessary to do the whole enchilada, which is boots on the ground."

In 2012, a largely Dutch-led group went public with another Mars plan, this one to colonize the planet with men and women who would agree to travel there and never return. The group, a nonprofit foundation, does not yet have the technical expertise to mount such an effort, but it certainly caught the public's attention and imagination when it put out a call for volunteers. In September 2013, Mars One released this statement:

"The first round of the Mars One Astronaut Selection Program has now closed for applications. In the 5 month application period, Mars One received interest from 202,586 people from around the world, wanting to be amongst the first human settlers on Mars."

Applications came in from 140 nations, with the most interest from the United States, India, and China. The Mars One plan calls for a seven-year training for six to ten groups of four people each. Then, if the technology and funding come together, they will go on their one-way trip to Mars.

Some will not doubt see the volunteers as unhinged and probably antisocial or suicidal. Some may be.

But there certainly are other ways of looking at their interest and fascination with Mars: People are eager to go, or see others go, where no man or woman has gone before. Some dream of conducting the most exciting science ever undertaken, or at least learning about it. Others can't wait to be part of the greatest technological endeavor in human history. Still others want to pioneer the push to make humans a multiplanetary species.

And some want to be part of a larger-than-life effort even if they never get off the planet. The group HI-SEAS—which is led by Cornell University and the University of Hawaii at Manoa— has begun running four-, eight-, and twelve-month isolation studies on the northern slope of Mauna Loa, in a lava field.

Six people at a time remain in a geodesic dome simulating life on Mars, with only brief trips outside (in a space suit, of course). The program, funded by NASA, is primarily a study about food and cooking, which have not gotten high marks in long-duration stays on Earth and in space. But it's also about living together in close quarters, which space travel and potential colony life on Mars require. More than 700 people applied for the first HI-SEAS venture.

The Mars Society is also preparing for a one-year isolation mission in the Canadian Arctic, a less clement but more Mars-like environment for sure.

Former NASA astronaut Buzz Aldrin, the second man to step on the moon, has become a vocal advocate for sending humans to Mars, and has written a book explaining why. Not surprisingly, he sees Mars travel in terms consistent with being a lunar pioneer.

The Mars500 project, sponsored by Russia's Institute of Biomedical Problems, has been simulating Mars exploration conditions by keeping scientist-astronauts in a small steel container for long periods to test their health and reactions. Here a Mars500 project crew member from the European Space Agency walks on a terrain simulated to replicate that of Mars.

"The people who go there will be remembered in history as pioneers and great men and women," he said. "The world leader who makes a commitment to establishing a permanent presence on another planet will also be remembered as a pioneer and hero for hundreds of thousands of years. This is what greatness is all about."

GOING TO MARS WITH SPACEX

And then there's Elon Musk, the founder of Space Exploration Technologies (SpaceX) and the electric car company Tesla Motors and a man of endlessly grand ideas. Mars has been on his radar screen for a long time.

The founding of SpaceX, in fact, flowed to some extent from Musk's interest in delivering a terrarium garden to Mars to test if people could live off the land. This was after Musk had become a wealthy man from selling the Internet purchasing company PayPal, and so he had funds to quite possibly pay for the experiment. But he quickly learned there was no spacecraft available to deliver him or his experiment, and in time he decided to build his own.

Since then, SpaceX has sent numerous rockets and capsules into space, including cargo spaceships that have supplied the International Space Station. Tesla has also become a thriving business, so Musk has a track record of thinking big and then, to the surprise of skeptics, delivering.

With this in mind, Musk's vision for sending humans to Mars doesn't involve a small landing party—"a few flags and footprints"—but rather the creation of a substantial, self-sustaining colony by 2100. He envisions a "New World," rather like the American colonies of the earliest days, where people could go to start new lives, build new towns, dig iron ore out of the ground, set up pizza shops.

He sees the effort as one involving NASA, other national space agencies, other space companies, with SpaceX as the catalyst. The costs would be enormous, but Musk has made his name by driving down prices for space launches and electric cars.

His plan requires a major conceptual leap—flying reusable spacecraft more like a Disney cruise ship or a jumbo jet that can carry hundreds of people to Mars. The

The Mars Desert Research Station sits in the rocky, red desert about 40 miles outside Hanksville, Utah. Scientists simulate life on Mars as accurately as possible at the base—dressing in space suits and living in the cramped hut for weeks at a time.

>> Buzz Aldrin—aeronautical engineer, astronaut, and the second man to walk on the moon—is a strong advocate for sending explorers and, ultimately, settlers to Mars.

ship wouldn't come and go from Earth, but would meet passengers at a low Earth orbit transfer station.

Then there would have to be an infrastructure on Mars that would be both safe and appealing. Becoming a multiplanet species is not only important for human evolution—maybe even essential—but it should also be "interesting, exciting, fun." Musk feels strongly enough about the idea that he says he's willing to someday put much of his substantial personal wealth into it—assuming, of course, that enough of the public is also interested in the adventure.

There's an urgency to Musk's vision, and SpaceX is developing a superheavy lift rocket that could send the heaviest payloads ever into space. He thinks the first exploratory trip could be possible in as little as 12 years.

"This is an opportunity opening up on Earth for the first time in four billion years, but we can't count on the opportunity remaining forever," he said. In his eyes, the biggest danger facing humanity is not an overloading of Earth—"I believe billions of people are likely to live on just fine for a long time"—but rather that our technological abilities and inspirations will eventually collapse, as they have in eras past.

"We know that the Egyptians built the pyramids but then lost that ability," he said. "The Romans built a vast system of aqueducts and created indoor heating and plumbing, but then

SpaceX founder and CEO Elon Musk stands by a Falcon 9 rocket, which launched his firm's Dragon capsules bringing cargo to the International Space Station. Travel to Mars will require an even more powerful launch rocket, and SpaceX is developing a reusable one powered by oxygen and methane.

those technologies were lost to Europe for a thousand years. And the Chinese once had the biggest fleet in the world, and then burned the whole thing.

"Civilizations can reach a high technological level and then stop. And now we're really all one civilization and arguably are at more risk that civilization will all decline at the same time.

"That's why it's so important to establish this colony while the window is open. Because it could quickly close."

THE RISKS OF RADIATION

Putting aside the inherent dangers of liftoff and landing, is it safe to go to Mars? With today's abilities to shield astronauts from hazards and meet their basic needs, the answer would have to be a pretty definite "no." It's not that Mars is not inherently more dangerous than the moon or other potential extraterrestrial landing sites. Location is the issue—a three-day trip to the moon versus a six-month one-way trip to Mars. Or, as some portray it, the difference is between swimming the English Channel and swimming the Atlantic Ocean.

And during that long swim across the Atlantic, a Mars-bound crew would be constantly showered by radiation. These high-energy, highly penetrating particles are similar to those produced by power plants and increase risks for future cancers.

But there are important differences, and a long-term exposure to radiation in deep space would be an entirely new experience for humankind. This is because Earth's magnetic field and atmosphere protect us from the kind of radiation that would assault Mars-bound astronauts—largely galactic cosmic rays and solar-particle events.

The solar events are far more dangerous, but also far less prevalent. They occur when solar flares and coronal mass ejections shoot very high-energy protons out into the solar system. If an astronaut is not in a heavily shielded area during a solar storm, he or she could receive a damaging or even fatal dose of the radiation.

Galactic cosmic radiation is primarily a by-product of exploding stars (or stars going supernova), which send out high-energy particles capable of penetrating today's spaceships. Since stars explode throughout the galaxy, cosmic rays are ubiquitous in space.

Galactic cosmic rays can be blocked to some limited extent by the hydrogen in water or a thick shielding around a spaceship. The protection, however, will inevitably be incomplete because the spaceship can take on only so much added weight. Researchers have studied and modeled the likely radiation exposure during a trip into deep space for some time. But Curiosity's RAD instrument has provided the first actual measurements, and has done so in a number of ways.

Donald Hassler is a space weather expert and research director at the Southwest Research Institute, which developed the instrument with the help of the German national space research center. The instrument was designed to measure radiation in the Martian atmosphere and on the surface, but its most important tasks clearly

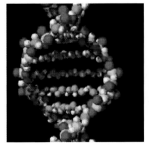

BLASTED BY RADIATION Solar and cosmic rays remain a top challenge in Mars exploration, since radiation damages DNA at its molecular level.

MORE THAN CURIOSITY >> Artists' representations
of the Mars Express Orbiter from the European
Space Agency (above left) and Mars Obiter
Mission spacecraft (right) from the Indian Space
Research Organization, currently in Mars' orbit

involve human exploration. Indeed, RAD was the only instrument on Curiosity funded by NASA's Exploration Systems Mission Directorate, along with Germany's national aerospace research center.

The RAD instrument was tucked deep inside the spacecraft during flight and had what is considered to be about as much radiation protection as an astronaut would have. Not only was it new to have that more realistic reading, but RAD was also able to break down and identify the different forms of radiation and measure them. Using its cosmic particle telescope, RAD could distinguish the galactic cosmic radiation from the solar-particle events, the direct heavy particle penetration versus the secondary radiation created when rays hit a barrier.

"We're the first instrument on the surface of another planet to measure radiation," Hassler said. "That was our designed mission, but we realized a year or so before launch that we could keep the instrument on the entire way to Mars. Measurements have been made before, but the monitors were always outside the capsule so they didn't measure how radiation would actually affect the astronauts. Now we know much better."

The results were not unexpected, but still sobering.

The daily dose in deep space was 1.8 millisieverts, which amounted to an exposure of some 455 millisieverts during the Curiosity trip. Even with that, the round-trip radiation dose in transit alone would bump up against the allowable increased risk from radiation, exposing astronauts to 100 times more radiation than on Earth.

RETHINKING RISKS

The radiation measurements continued on the surface of Mars and came up with considerably lower exposure rates—about one third of the incoming radiation measured in deep space. But combined with the round-trip flights, any substantial time on Mars would clearly put an astronaut well over the NASA allowed limit—which is a 3 percent increase in the likelihood of developing a radiation-induced cancer.

Nonetheless, the Mars Science Lab/Curiosity radiation numbers were not considered showstoppers. Many factors could change the equation, ranging from medicines NASA hopes will protect astronauts from radiation, to more imaginative shielding, and new propulsion systems that could speed astronauts to Mars. There's a push on now as well to change that official level of allowable radiation risk set by NASA.

Taber MacCallum is an organizer of the Inspiration Mars project, which is aiming for that 2018 flyby. While the project is formally a philanthropy, the group is working closely with NASA and would have to follow the agency's exposure guidelines. But MacCallum, a co-founder of the Paragon Space Development Corporation, believes the current guidelines are unrealistic and harmful.

"Let's get real here," he said. "People make choices all the time that increase their risks of disease and death much more than three percent. They climb Mount Everest, they smoke, they join the military. For something like traveling to Mars, something which would have enormous benefits for the country, a little more risk seems acceptable. Goodness knows, there are lots of people eager to take the risk."

Still, the Inspiration Mars effort—which is being funded initially by financial adviser and space enthusiast Dennis Tito—has worked out a way around the 3 percent limit. Its plan is to send a couple in which both astronauts are over 50 years old, a plan that reduces their risk of radiation-induced cancer because their life expectancy would be much shorter than a 30-year-old's.

A recent study by Tore Straume, a radiation biophysicist at NASA's Ames Research Center, reflects the serious, if limited, official interest in understanding the full range of life challenges men and women would face during long-duration space travel and even a colonizing of Mars. Straume found, not surprisingly, that radiation in space would have harmful effects on pregnant women and any fetus they would be carrying. Men's sperm would also likely be damaged in deep space, other researchers have found.

These issues could be avoided on a trip to Mars, but they are central to the notion of starting a human colony on the planet. To survive and grow, a colony would need children to be born. And it's unclear now whether that will ever be possible.

An artist's rendering of a future greenhouse suggests one possibility for self-sustenance on Mars.

<< A jubilant Dennis Tito lands in Kazakhstan after his nearly eight-day trip to the International Space Station. An American engineer and entrepreneur, Tito is now planning Inspiration Mars, a round-trip venture for two astronauts to fly to Mars and back by 2018.

To those won over by the lure of Mars, the undeniable and large obstacles are often seen as challenges, not reasons to stop the effort. JPL's rover-landing mastermind Adam Steltzner, for instance, said that while sending astronauts to the surface of Mars in the near future isn't feasible, it certainly could be a bit further out. What's needed, he added, is a national decision to make such a landing a priority—something akin to President Kennedy's call to send a man to the moon. Once that happens, Steltzner says, the engineers will do the rest.

And when it comes to the risks inherent in traveling to Mars, this country has some experience with that as well. After all, when America was the "New World," quite a few aspiring colonists and some entire communities didn't make it across the Atlantic, or perished not long after landing. But those who did establish a new life here generally did quite well for themselves.

Elon Musk's take on the risks and rewards of Mars is perhaps instructive. He very much wants to travel to Mars and says it has long been one of his life's dreams. It's the adventure of all adventures. In fact, he even wants to die there. He just wants to make sure that will happen naturally, not in a spaceship landing gone awry. That, he firmly believes, is an achievable goal.

MARS CIRCA 2100

<< BUILDING CAMP GALE

Today it's a dream. But sending humans to Mars will become a reality if spacefaring pioneers have their way. This artist's rendering shows the Peace Vallis area in 2100, when humans have taken the place of rovers.

REFLECTIONS

CHAPTER 14

Mars is so distant and alien, but suddenly familiar and knowable as well

IF YOU LET IT, Mars tugs at you. There's a gravitational force that pulls the imagination. Like gravity, it's weak in small doses. But as the proximity increases, so does the pull.

Going into this project, I thought the thrill and discovery would center around how different Mars is from Earth. The parched, red landscape that has generally defined Mars for years, the frigid and irradiated surface, the atmosphere that barely exists. We've all been trained to think of Mars as alien and different and hostile, what with all those angry Mars gods and space invaders with large green heads or very long legs. It was always puzzling to me that many people were pushing to send astronauts to Mars. The thought that generally came into my mind was: Why?

But the surprise, and the reality, is quite the opposite of the conventional wisdom about Mars. It surely is a forbidding place now, but we're just seeing it at one point in time. Like Earth, and most everything else in our universe, Mars had evolved. Not in the direction of Earth, but still it underwent dramatic changes that ended with a parched and frigid planet. It may well have started, however, as a considerably wetter and warmer one.

What this means is that we can now say with some confidence that had a spaceship flown to Mars some four billion years ago, there's good reason to think the planet it landed on would have looked and operated rather like

Earth—that is, an Earth just before or very early in the commencement of its most primitive life. We know that both planets formed at about the same time—4.5 billion years ago—and both are in what is considered the habitable zone of their sun, where it's not too hot or too cold for water to sometimes flow as a liquid.

After cooling down from its steamy birth and enduring a long period of heavy meteorite bombardment, Mars—like Earth—settled down. It had flowing rivers and standing lakes, was enveloped by a relatively thick atmosphere, protected by a surrounding magnetic field, and, as the Curiosity scientists concluded, was entirely habitable in some places. It was not quite an Earth twin (too far from the sun for that), but it was at least an Earth cousin.

So coming to terms with Mars involves, as much as anything, coming to terms with time. To me, that is the gift of the Curiosity mission. Its findings have documented and made clear as never before the transformation of another planet over billions of years, changes as dramatic as those on Earth but with a very different rollout and endpoint. That's an understanding that invites you to reflect on astronomical time—the parade of eons that have ended with us, here now on one among the billions and billions of planets in the universe.

Think of the first powdered rock sample pulled out of Yellowknife Bay. It wasn't reddish colored like the surface today, but rather was a grayish green or grayish blue. The red planet, it turns out, once wasn't red at all. Today's red is but a stage in the planet's evolution, one where iron oxide (rust) dominates. Long ago, Mars was no more reddish than Earth.

Similar vast changes have occurred on Earth, but we take them for granted. Oxygen, we know, is essential for most life, yet it is actually quite a toxic substance. Early and middle Earth had little oxygen, which was good for the organisms alive at the time. But the big explosion of life on our planet took place after organisms began to adapt to the presence of the oxygen that was being produced in increasing amounts by photosynthesizing bacteria in the oceans.

So like living things, planets evolve. They can evolve toward greater complexity as on Earth, or they can lose complexity. That certainly seems to have happened on Mars. In geological or astronomical time, however, such a loss is not

FORWARD TO THE PAST >> The Mars terrain is parched and cold, but it no longer seems quite so desolate. Curiosity has revealed an ancient past when the planet was capable of holding the water needed to form rocky outcrops that look somewhat similar to formations on Earth. Some imagine a day when Mars can be terra-formed (sequence at right): changed from its harsh present to a more Earth-like future that just might be similar to its past.

SW

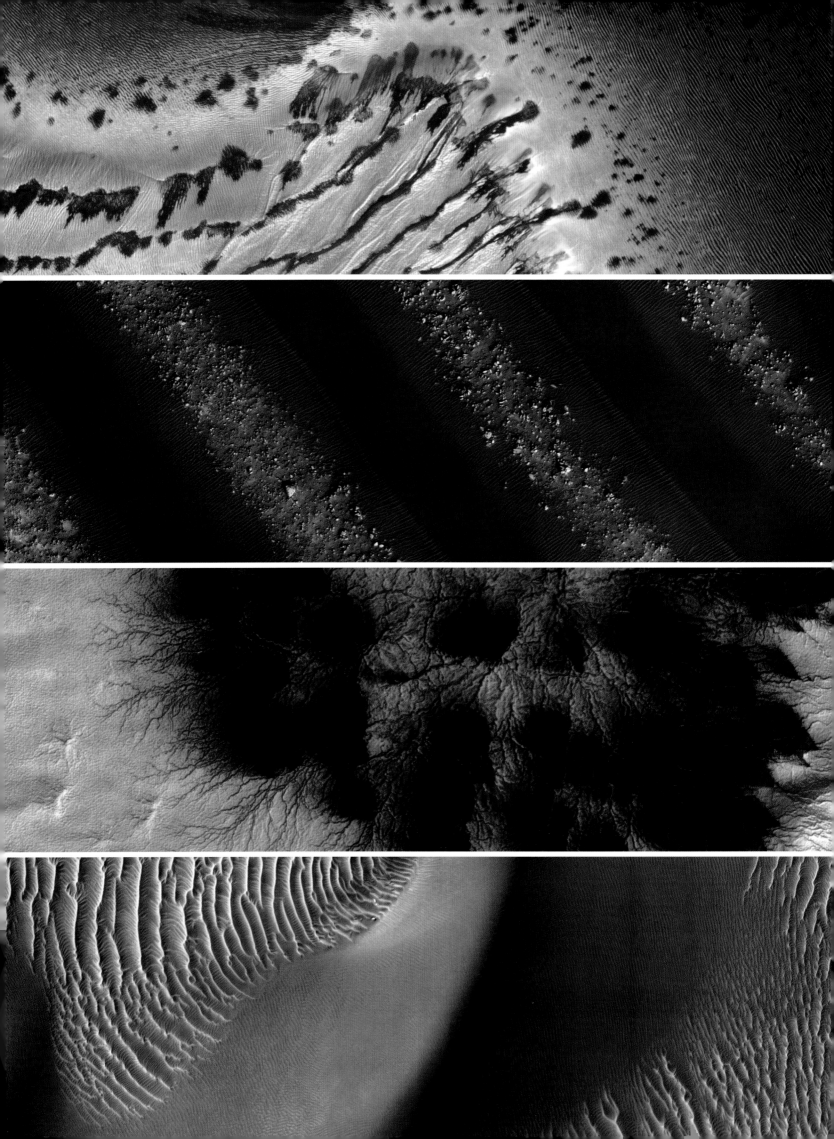

necessarily of great moment. What's important is that the greater complexity once was there. And on Mars, that means understanding those early days when the planet was clearly habitable—a stage set for players, however small and primitive, to act. A world of possibilities.

The dramatic changes that befell the planet quite early made it nearly impossible for any of those potential players to evolve into creatures more complex than single-cell microbes. But from the point of view of the search for life beyond Earth, that too hardly matters. While coming across the remains of a more evolved creature on Mars would be hugely exciting, its complexity would not be the primary breakthrough; evolution, after all, is pretty well understood.

It's the origin of life, the moment (or moments) of genesis, that remain the great mystery. And if a genesis occurred on Mars as well as on Earth, then something profound has changed in our understanding of the universe.

Remember all those hundreds of billions of planets circling hundreds of billions of other stars? Many of them are orbiting in their solar systems' habitable zone, where water can be liquid. And all of them were showered from space by the same organic material and other compounds needed for life.

If life arose twice (on Earth and Mars) in a single solar system, one that as far as we know is not at all unique, then the likelihood grows that life began on many, many planets. We're not going to get to any of those faraway places anytime soon, but we are on Mars and engaged in a decades-long search for signs of life there. Given the enormous implications of a discovery that life once existed on Mars, or perhaps still does deep underground, the stakes and possible rewards are awfully high.

A CLOSE RELATIVE

The second enticing possibility is that Mars will someday help us to answer not only questions about life beyond Earth but also about life *on* Earth—most especially how life here might have begun.

Swaths of Mars have been covered over by lava from massive, long-ago volcano flows, but a significant proportion of the planet remains untouched by those kinds of changes and is rather like it was four billion years ago. Brush away that covering of red dust and dig past the meter likely modified by cosmic and solar radiation, and you have something close to Mars as it was when life was beginning on Earth (and perhaps on Mars, too).

On Earth, that geological record has long since disappeared. The churning of plate tectonics takes surface layers and pulls them down into the cauldron below the surface before spitting them out again. The rocks, minerals, and organics may ultimately return to the surface, but they will have been forever changed. The conditions on Earth that gave rise to life some 3.8 to 4 billion years ago will never be replicated, never really understood.

There have been few, if any, signs of similar plate tectonics remaking Mars. Rock is to some limited extent recycled through volcanoes, but the surface is not pulled under and transformed.

<< The drama of the geological past of Mars shows in the variety of textures found across the landscape. From top to bottom are the Russell Crater Dunes, a crater in Noachis Terra, Mars's seasonal cap of dry ice, and Proctor Crater.

What's more, Mars has not been made over by life as Earth has. Living things interact with and change the rocks, the water, the atmosphere, and so also hide the conditions present when life began on Earth some 3.8 billion to 4 billion years ago. Even if Mars did once have life, it was simple and most likely limited in its range. So it never got a chance to terraform the surface, as happened on Earth.

It is not surprising, then, that a number of scientists on the Curiosity team actually specialize in understanding early Earth. They also specialize in understanding planetary change generally.

It's a perhaps peculiar thought, but some day discoveries about conditions on early Mars may help us understand how life began here.

They're called blue-berries: round balls, up to three millimeters in diameter, made of the iron oxide mineral hematite. Discovered by the rover Opportunity, they may look like life-forms, but they were most likely shaped by geological forces.

>> Activity in gullies such as new deposits are a priority focus for HiRISE. The new deposits in Gasa Crater are distinctively blue when the images are color-enhanced.

And then there's the question of panspermia—whether life on Earth may have actually come from Mars.

The hypothesis has been around for some time, but recently received some buttressing. The source was Steven Benner, a highly regarded and iconoclastic origins-of-life chemist with the Westheimer Institute of Science and Technology in Florida, who was speaking at an annual conference of geochemists in Italy.

He has been struggling, along with others active in the field of "synthetic life," to create life from nonliving parts. Scientists have made progress in understanding the logic of this dawning, but they continue to face enormous hurdles.

One is that the ribose in RNA—which many scientists have concluded played a central role in the rise of the earliest life—falls apart when you try to build it in water. Yet water is also assumed to be essential for life. And single-strand RNA is assumed to be essential to the later formation of double-helix DNA. It's a major conundrum for Benner and others in the field.

One addition that keeps the water from breaking the ribose apart, Benner has found over years of study, is the presence of a form of the element boron. While geologists say boron was too scarce on early Earth to support any widespread creation of RNA, it was apparently more abundant on early Mars. One sign of its presence there came recently in the analysis of a meteorite that delivered some Martian boron to Earth.

Benner has also found in his lab that if a form of the element molybdenum is added to the mix, the boron-steadied compounds are rearranged to form a stable version of ribose—the R in RNA. Again, the element was far more available on early Mars than early Earth.

So the question arises: Did RNA on Mars lead to actual DNA-based life? And did those life-forms then travel to Earth on rocks kicked up when a meteorite struck Mars?

"Basically, we went looking on Mars because the origins-of-life options on Earth just aren't looking very good," Benner said. "It seems to me the evidence is leading to the real possibility that we are actually all Martians; that a rock kicked up by a meteorite brought life to Earth from Mars."

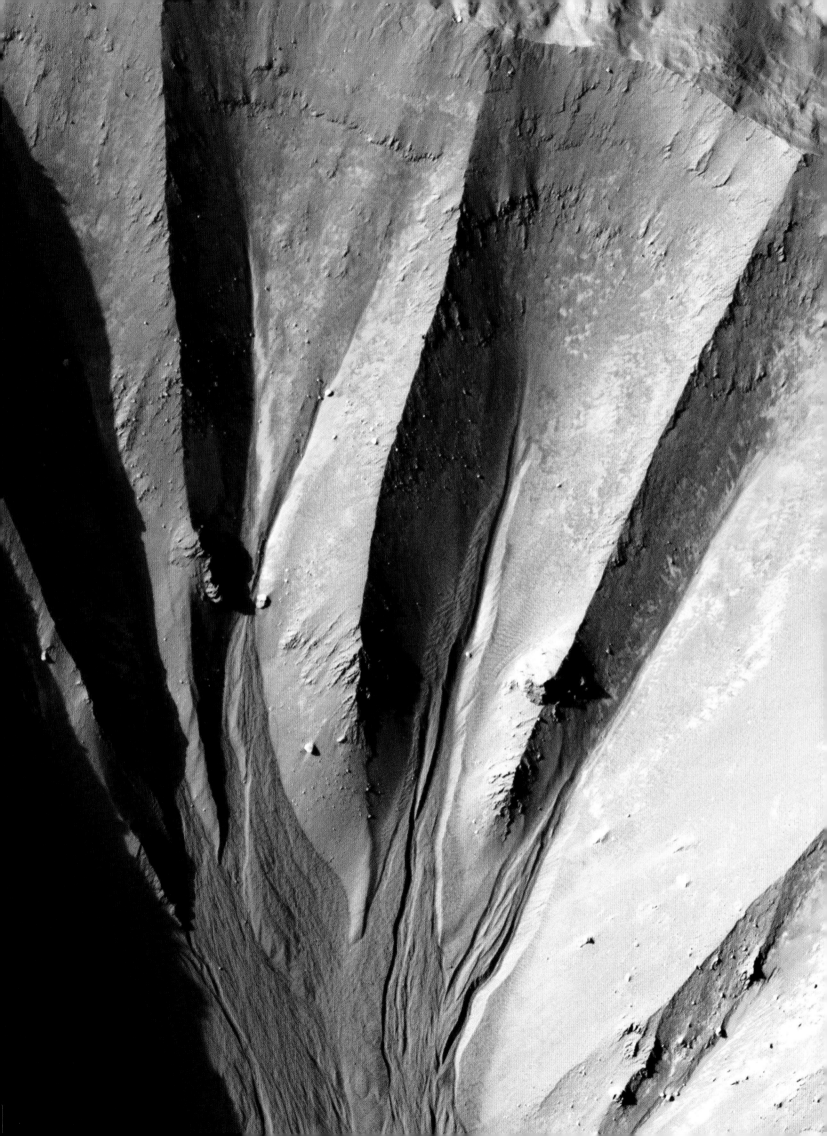

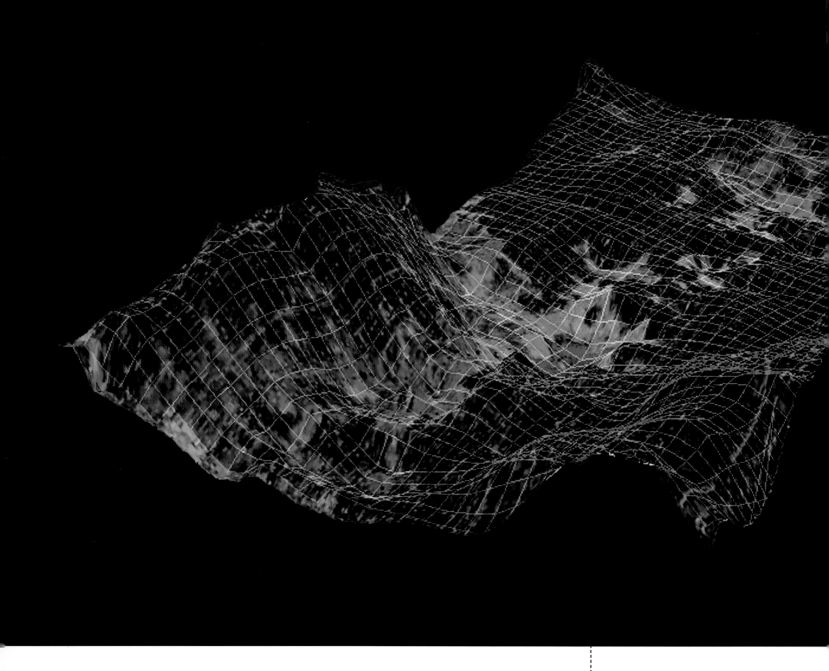

Our cousin planet just may, after all, be our mother planet as well.

For me, at least, the scientific rationale for exploring and eventually sending astronauts to Mars was getting ever stronger. The scientific questions to address don't get any larger or compelling. As JPL's Gentry Lee put it, finding confirmable signs of ancient (or more recent) life on Mars would be among the handful of most important discoveries ever, and perhaps the single most important ever. That's one man's opinion, but it's pretty hard to argue with.

A related question is equally important: What happened to Mars in its distant past to transform a place that was wetter and warmer into one that was much drier and colder?

Theories so far have focused on the long-ago loss of much of the Martian magnetic field that, in effect, held the atmosphere in place and allowed for warmer and wetter conditions. Remnants of a magnetic field are found in magnetized Martian rocks, almost all of which are in the planet's less altered southern hemisphere.

Mars is about half the diameter of Earth with a much smaller core—the molten mix of iron and other elements that creates and supports the magnetic field that surrounds some planets. That means to scientists that the magnetic field was always weaker than on Earth. Did it just

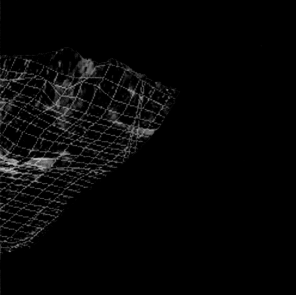

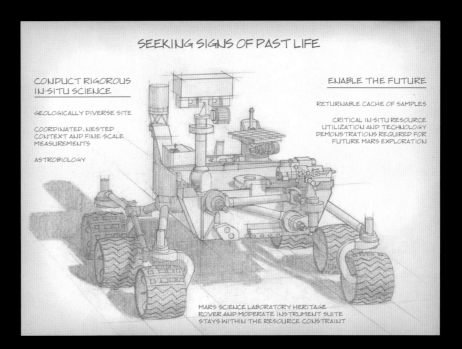

SEEKING SIGNS OF PAST LIFE

CONDUCT RIGOROUS IN-SITU SCIENCE

GEOLOGICALLY DIVERSE SITE

COORDINATED, NESTED CONTEXT AND FINE-SCALE MEASUREMENTS

ASTROBIOLOGY

ENABLE THE FUTURE

RETURNABLE CACHE OF SAMPLES

CRITICAL IN-SITU RESOURCE UTILIZATION AND TECHNOLOGY DEMONSTRATIONS REQUIRED FOR FUTURE MARS EXPLORATION

MARS SCIENCE LABORATORY HERITAGE ROVER AND MODERATE INSTRUMENT SUITE STAYS WITHIN THE RESOURCE CONSTRAINT

FUTURE PLANS >> NASA is already drawing up plans (above) for the launch of a Curiosity 2, set for 2020. It will rely on the basic technology of the current rover, thereby making it less expensive to build, but it will deploy different instruments. Scientists also want to send a mission to Mars that will begin collecting rocks to bring back to Earth (left). Meanwhile, orbiting satellites continue to determine with increasing precision the contours, mineral makeup, and thermal properties of the Martian surface (opposite).

peter out? Did one or more large meteorites smash into Mars and set in motion a process that destabilized the core and broke up the magnetic field? Was it something else entirely?

All that can be said with real confidence is that dramatic change came to early Mars making it increasingly more dry and cold. We, on the other hand, have had a stable planet that kept its magnetic field, kept its atmosphere, and kept its conduciveness to life. We've been lucky.

THE HUMAN FACTOR

Just after the successful launch of the rocket that would take MSL and Curiosity to Mars, I received a very excited call from friend and SAM deputy principal investigator Pan Conrad to spread the news. She had told me several years before that she was preparing for the mission as if it was the Olympics, and she was in professional and physical training. Now the Olympics had started, and it was hard to take in—like any huge and long-anticipated event. Let the games begin!

I would see Pan periodically at Goddard, where she works, and at JPL, where she chaired some of the science working-group meetings. We would talk about the mission, and I asked if it was really like the Olympics and whether she had trained enough.

Her answer: It was better than the Olympics because it lasted so long. Not always at the same fever pitch of excitement, but often enough to make it thrilling. And was she prepared for what turned out to be a very long marathon run? No, she says. She should have trained even more.

I often think of Pan's enthusiasm and dedication when I speak with Curiosity engineers, scientists, and managers. The belief is widespread that they are privileged to be involved, that the Curiosity mission is an undertaking of significance that may come but once in their lives. They feel responsible for its success, and so many give it their all.

There are, of course, periodic conflicts within the team, some second-guessing and frustrations. Was the sky-crane maneuver overengineered? Should Curiosity have gone straight to Mount Sharp rather than taking the Yellowknife detour? Will SAM ever come up with a definitive finding about organics? Why hasn't the rover been able to drive as fast as forecast and planned?

As the months pass, some participants move on to other work or return to a more normal life where the Mars Science Olympics don't loom quite so large. But doubtless most will take back to their jobs, families, and communities an expanded view of what's possible based on being part of such an enormous and fruitful undertaking.

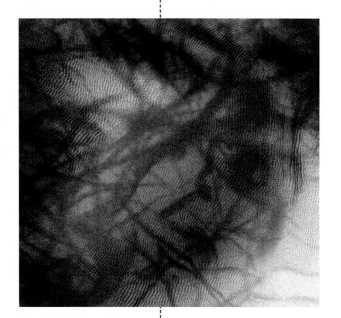

Dust devils on Mars expose underlying darker material as they travel across a dune field. The result is a criss-cross of trails interspersed with striking swirls.

The avalanche of science papers attests to the size and significance of the effort. By the end of 2013, the prestigious journal *Science* had published 16 Curiosity papers.

This is how the story of our cousin (mother?) planet generally comes out, in what may seem like obscure and confusing snippets. But the plotline is pretty clear and appealing: Mars was once quite different, and a lot more like Earth.

But the mission doesn't only highlight what's compelling about Mars. It also makes clear that our exploration and parsing of the planet has just begun. The unanswered questions vastly outnumber the ones nailed down with some certainty. And Curiosity, for all its capabilities, cannot even begin to address many of the most intriguing and important mysteries. So more Mars mission are necessary if we really want to excavate, expose, and understand the planet's past.

Mars missions, however, are expensive. Getting Curiosity to Gale Crater and operating it there for almost two years cost about $2.6 billion.

The next few missions to come are far less expensive, but the technological advances and workforce needed to bring a Mars rock back to Earth or ultimately send astronauts to the planet will make the bill for the Curiosity expedition look bargain-basement. Musk and other space entrepreneurs are convinced they can make the trip cost far less—maybe $500,000 per person for a round-trip—but that is a goal for the quite distant future.

A PLAN FOR MARS >> The components of the Mars Transit Vehicle launch into Earth orbit, where the assembly crew hands the reins over to the Mars One crew before returning to Earth. After a final check of systems the Transit Vehicle is launched on a Mars Transit Trajectory. This is the point of no return; the crew is now bound to a 210-day flight to Mars.

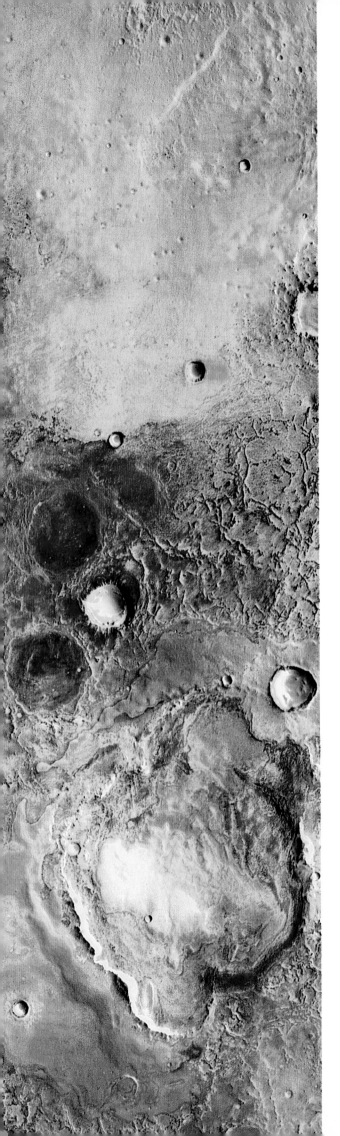

« A satellite view of Mars's Meridiani region
evokes a passage from Ray Bradbury's *The Martian
Chronicles*: "You know what Mars is? It's like a thing
I got for Christmas seventy years ago—don't know if
you ever had one—they called them kaleidoscopes,
bits of crystal and cloth and beads and pretty junk.
You held it up to the sunlight and looked in through
at it, and it took your breath away. All the patterns!
Well, that's Mars. Enjoy it."

The inevitable question that arises is whether a long-term program to explore Mars and send astronauts there is worth the many tens of billions it will cost, especially when taxpayers are footing a good part of the bill.

Many people clearly don't share the sense of excitement that NASA and its fans feel about space travel and exploration, and so they are reluctant to pay for it. Earth, after all, has its share of big problems.

Travel to Mars is also inherently risky, much riskier than anything done in space so far. And that's before any humans are launched in its direction. Is Mars worth all the effort?

My time embedded with the Curiosity mission has prejudiced me in favor. The feeling is palpable at JPL: Something extraordinary is under way, an endeavor where the never ending challenge and burden of getting so many things right is routinely embraced because of the size of the prize.

What the public ultimately concludes about the value of unraveling the story of Mars and sending people there means a lot; the effort, in fact, really can't succeed unless it becomes a national priority.

Curiosity plays a big role in this decision-making process. It has raised the bar both technically and scientifically; it has introduced the public to a new (very old) habitable Mars and provided a test drive of the dramas that bigger and some-day manned Mars missions would deliver. It has made many significant discoveries while demonstrating an American and international ability to operate an extremely complex robot machine in often hostile circumstances very far away.

For me, Curiosity has forever increased that gravitational pull of interest about our cousin planet. We've been connected since the start, both born from the same disk of debris orbiting our protosun. It's not inevitable that our futures will bring us closer again, but it's quite possible, and quite desirable, too. A challenge, a prize, a time capsule like no other, Mars beckons.

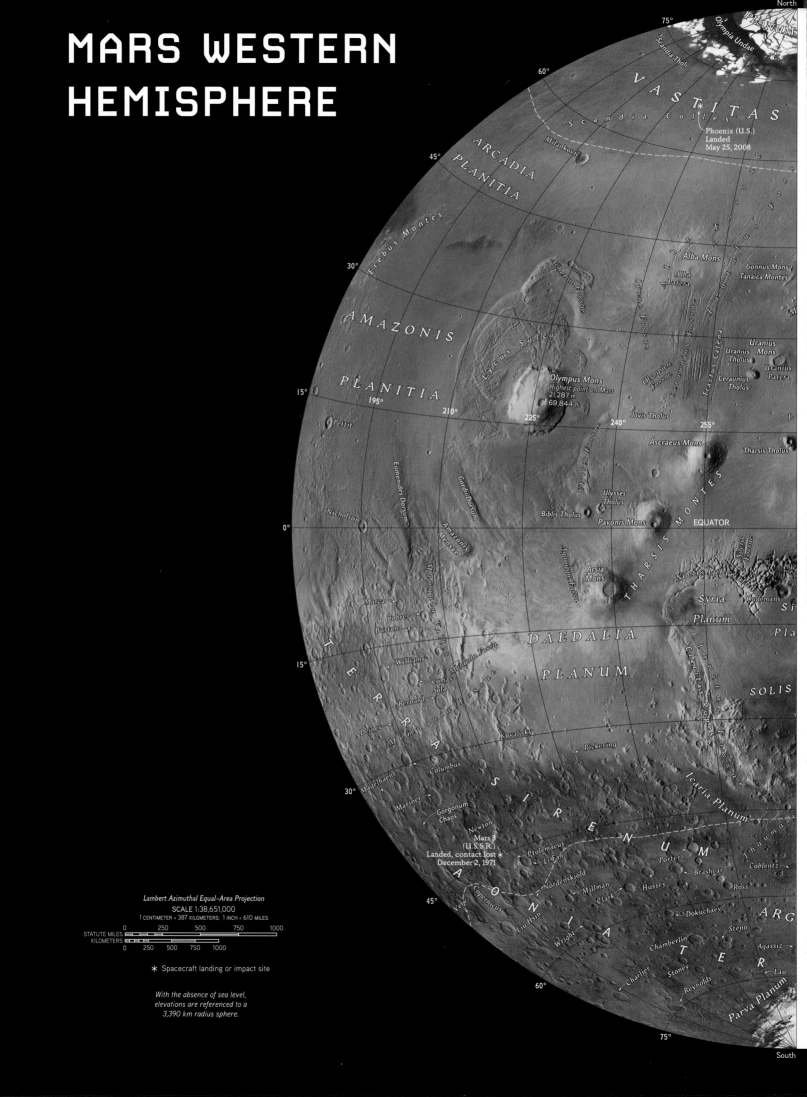

MARS WESTERN HEMISPHERE

Lambert Azimuthal Equal-Area Projection

SCALE 1:38,651,000

1 CENTIMETER = 387 KILOMETERS; 1 INCH = 610 MILES

0   250   500   750   1000

STATUTE MILES

KILOMETERS

0   250   500   750   1000

✳ Spacecraft landing or impact site

With the absence of sea level,
elevations are referenced to a
3,390 km radius sphere.

South

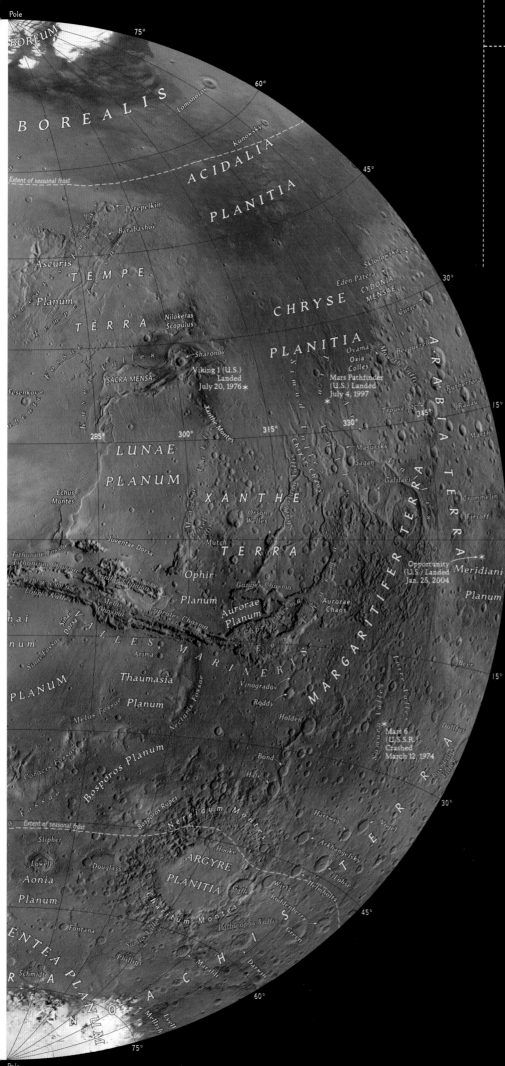

Mars has the highest volcano in the solar system, 16-mile-high Mount Olympus. This side of the planet also features Valles Marineris, a rift canyon 2,500 miles long, 120 miles wide, and up to 23,000 feet deep.

MARS EASTERN HEMISPHERE

North

75°
60°
45°
30°
15°
0°
15°
30°
45°
60°
75°

South

VASTITAS

PLANUM

Extent of

Micoud
Ivol

Ismenia
Patera

DEUTERONILUS
MENSAE

PROTONILUS MENSAE

Colles Nili

Renaudot

NILOSYRTIS MENSAE

Astapus Colles

Peridier

ISID

ARABIA TERRA

Maggini

Cerulli

Moreux

Rudaux

Quenisset

Perce Fossae

Nili Fossae

Arena Colles

SYRTIS

PLANI

Luzin

Flammarion

Baldet

Antoniadi

MAJOR

Beagle 2 (U.K.)
Crashed ✳
December 25, 2003

Cassini

Schöner

Pasteur

Gill

Tikhonravoy

PLANUM

Du Marth

Henry
Arago

Capen

TERRA

Schroeter

Libya

Targi Valles

Janssen

Teisserenc
de Bort

Fournier

Jarry-Desloges

Oenotria Plana

Schiaparelli

Meridiani

Planum

Pollack

Dawes

Oenotria Scopulus

Huygens

TYRR

Madler

SABAEA

Saheki

Millochau

Harris

TER

Flaugergues

Denning

Bouguer

Cankuzo

Schaeberle

Terby

Hadriacus
Palus

Marath
Wislicenus

Niesten

Bakhuysen

Newcomb

Lowest point on Mars
-8,180 m
-26,838 ft

Alpheus
Colles

HELLAS PLANITIA

Dao

NOACHIS TERRA

Le Verrier

Rabe

Hellespontus Montes

Hellas Chaos

✳ Mars 2
(U.S.S.R.) Crashed
Nov. 27, 1971

Kaiser

Proctor

Amphitrites
Patera

Maunder

Russell

Barnard

Malea Planum

PR

Mitchel

Dorsa Brevia

Gilbert

Pityusa
Patera

Holmes

Sisyphi
Planum

Promethe

Prometh

PLANUM A

Lambert Azimuthal Equal-Area Projection
SCALE 1:38,651,000
1 CENTIMETER = 387 KILOMETERS; 1 INCH = 610 MILES

0   250   500   750   1000
STATUTE MILES
KILOMETERS
0   250   500   750   1000

✳ Spacecraft landing or impact site

*With the absence of sea level,
elevations are referenced to a
3,390 km radius sphere.*

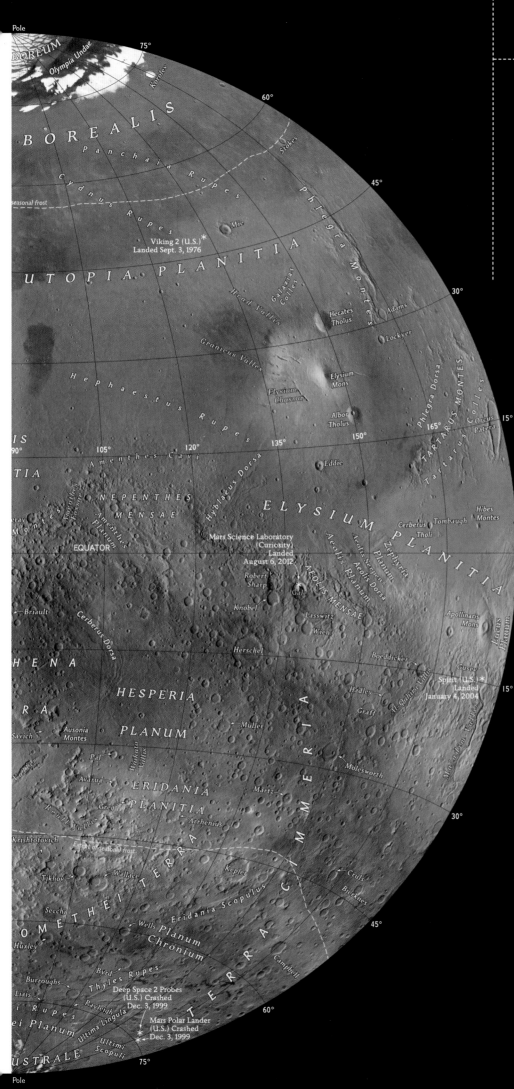

The northern hemisphere of
Mars is generally a flat plain,
and considerably lower than the
south. The crater count in the
north is also lower than in the
south, implying that the northern
surface is younger.

ABOUT THE AUTHOR

A science writer and reporter on international affairs, Marc Kaufman has been a journalist for 35 years, writing primarily for the *Washington Post* and the *Philadelphia Inquirer*. His previous book, *First Contact: Scientific Breakthroughs in the Hunt for Life Beyond Earth,* was named a top nonfiction selection for 2011 by *Kirkus Reviews*. He often lectures on the topics of astrobiology and space exploration, and he has spoken at the National Book Festival on the Mall sponsored by the Library of Congress; at NASA headquarters; at the SETI Institute; at the Humans to Mars summit; and on NPR's *Fresh Air*. He lives in the Maryland suburbs of Washington, D.C.

ACKNOWLEDGMENTS

First and foremost, I would like to thank my wife and muse, Lynn Litterine, for her support and guidance.

Since the Curiosity mission is such a large team effort, gathering information and insights naturally introduced me to scores of scientists, engineers, and managers. Many were very generous with their time and trust, but I would like to give special thanks to John Grotzinger, Pan Conrad, Dan Glavin, Michael Ravine, Bill Dietrich, and Becky Williams. I am also indebted to Guy Webster, press officer at JPL and the gracious recipient of more of my phone calls and e-mails than any other.

I was very fortunate to have Susan Hitchcock as my editor. She did a superb job of pulling together the many aspects of the book and making them work together. Art Director Marty Ittner and Illustrations Editor Nancy Marion gave the book its unique look and feel, and I thank them greatly for that, too.

My reporting at JPL and elsewhere was helped substantially by a generous grant from the National Geographic Society Expeditions Council. Others who helped in important ways:

Magdy Bareh	Don Hassler	Douglas Ming
Steve Benner	Joel Horowitz	Michael Mischna
Jeff Biesiadecki	Kevin Hussey	Michael Mumma
David Blake	Peter Iliot	Tracy Neilson
Thomas Bristow	Louise Jandura	Marisa Palucis
Fred Calef	Nina Lanza	Mark Petrovich
Stephen Clifford	Stéphane Le Mouélic	Susan E. Poulton
Richard Cook	B. Gentry Lee	Melissa Rice
Brian Cooper	Laurie Leshin	Matthew Robinson
Joy Crisp	Daniel Limonadi	Kirsten Siebach
Jason Crusan	Taber MacCullum	Katie Stack
David DiBiase	Paul Mahaffy	Adam Steltzner
James Donaldson	Justin Maki	Dawn Sumner
Ken Edgett	Michael Malin	Vandi Tompkins
Bethany Ehlmann	Robert Manning	Jennifer Trosper
Jen Eigenbrode	David Martin	Ashwin Vasavada
Douglas Ellison	Alfred McEwen	Mike Watkins
Ken Farley	Veronica McGregor	Chris Webster
John Grant	Michael Meyer	Rich Welch
James Green	Ralph Milliken	Roger Wiens
John Grunsfeld		Cary Zeitlin

ILLUSTRATIONS CREDITS

Cover: Images by NASA/JPL-Caltech. Stitching by Andrew Bodrov, 360cities.net.

Front Matter: 2-3, NASA/JPL/University of Arizona; 4-5, NASA/JPL-Caltech/MSSS; 6, NASA/JPL-Caltech; 6-7, MSSS; 8, Courtesy SpaceX; 10-11, Stocktrek Images/ Getty Images; 12, AP Photo/Waco Tribune-Herald, Duane A. Laverty; 12-13, Courtesy SpaceX; 14-15, NASA/JPL-Caltech.

Chapter 1: 16, NASA/JPL-Caltech/ESA; 17, NASA/ JPL-Caltech; 18 (UP), NASA/JPL-Caltech/MSSS; 18 (CTR), NASA/Bill Ingalls; 18 (LO), NASA/JPL, Pioneer Aerospace; 19, NASA/JPL-Caltech/University of Arizona; 20-21, NASA/Frankie Martin; 21 (LE), NASA and Glenn Benson; 21 (CTR), NASA/JPL-Caltech; 21 (RT), NASA; 22, ESA/DLR/FU Berlin (G. Neukum); 24, Brian van der Brug/POOL/epa/Corbis; 24-25, Brian van der Brug/POOL/epa/Corbis; 26 (UP), NASA/JPL-Caltech; 26 (LO), NASA/JPL-Caltech/University of Arizona; 27 (LE), NASA/JPL-Caltech/Lockheed Martin; 27 (RT), NASA/JPL-Caltech; 29, MSSS; 30 (UP), Michael Nelson/epa/Corbis; 30 (LO), NASA/JPL-Caltech; 32-33, Oliver Uberti, NGM Staff; Art: Nick Kaloterakis. NG Maps. Source: NASA. National Geographic Creative.

Chapter 2: 34-35, NASA/JPL-Caltech; 36 (UP), NASA/ JPL-Caltech/MSSS; 36 (LO), NASA/JPL-Caltech; 37, ESA/DLR/FU Berlin (G. Neukum); 38 (UP), NASA/JPL/ University of Arizona; 38 (UPCTR), NASA/JPL/University of Arizona; 38 (LOCTR), NASA/JPL/University of Arizona; 38 (LO), NASA/JPL/University of Arizona; 40-41, NASA/ JPL-Caltech/MSSS; 41 (LE), NASA/JPL/University of Arizona; 41 (CTR), NASA, J. Bell (Cornell U.) and M. Wolff (SSI); Additional image processing and analy- sis support from: K. Noll and A. Lubenow (STScI); M. Hubbard (Cornell U.); R. Morris (NASA/JSC); P. James (U. Toledo); S. Lee (U. Colorado); T. Clancy, B. Whitney, and G. Videen (SSI); and Y. Shkuratov (Kharkov U.); 41 (RT), NASA/JPL-Caltech; 42, NASA/JPL-Caltech/MSSS; 44, Book jacket copyright by Knopf Doubleday, a division of Random House, Inc., from *The Martian Chronicles* by Ray Bradbury. Used by permission; 45 (UP), NASA/JPL-Caltech/Malin Space Science Systems; 45 (LOLE), NASA/JPL-Caltech/University of Arizona; 45 (LORT), NASA/JPL-Caltech; 46 (UP), NASA/JPL-Caltech/ MSSS; 46 (CTR), James P. Crutchfield, 2012; 46 (LO), Map compiled by Fred Calef, JPL-Caltech; William Dietrich, UC Berkeley; Lauren Edgar, Arizona State; John Grotzinger, Caltech; Nicolas Mangold, LPGN; Marisa Palucis, UC Berkeley; Melissa Rice, Caltech; Scott Rowland, U. of Hawai'i at Manoa; Katie Stack, Caltech; and Dawn Sumner, UC Davis. Reprinted from *Science*, Vol. 343, January 24, 2014. Copyright 2014 American Association for the Advancement of Science; 47, NASA/JPL-Caltech/MSSS; 48, NASA/JPL/GSFC; 49 (LE), NASA; 49 (RT), Google Earth; 50-51, NASA/JPL/MSSS; processings and mosaic: Olivier de Goursac, 2013; 51, NASA; 52 (UPLE), Mars (oil on canvas), Spranger, Bartholomaeus (1546-1611)/Private Collection/Photo © Christie's Images/The Bridgeman Art Library; 52 (UPRT), wikipedia/Reconstructed from several online sources by Joe Haythornthwaite; 52 (LOLE), Thuvia, Maid of Mars, A. C. McClurg & Co., Chicago, 1920; 52 (LORT), WorldPhotos/Alamy; 53 (UP), Hubble telescope image/Encyclopaedia Britannica/UIG/The Bridgeman Art Library; 53 (LOLE), The War of the Worlds, 2012 (oil on canvas), Barry, Jonathan (Contemporary Artist)/ Private Collection/The Bridgeman Art Library; 53 (LORT), Popperfoto/Getty Images.

Chapter 3: 54-55, MSSS; 56, Mark Thiessen/National Geographic Creative; 57, NASA, JPL-Caltech, MSSS- Panorama by Andrew Bodrov; 59 (UP), NASA/ JPL-Caltech/MSSS; 59 (CTR), AP Photo/Damian Dovarganes; 59 (LO), NASA/JPL-Caltech; 60 (UP), NASA/JPL-Caltech/Malin Space Science Systems; 60 (LO), NASA/JPL-Caltech/MSSS; 61, AP Photo/Los Angeles Times, Brian van der Brug, Pool; 62, NASA/ JPL-Caltech/Malin Space Science Systems; 63, Purestock/Getty Images; 64 (UP), NASA/JPL-Caltech/ LANL; 64 (LOLE), NASA/JPL-Caltech; 64 (LORT), NASA/ JPL-Caltech/LANL/CNES/IRAP; 65, NASA/JPL-Caltech/ LANL; 66 (LE), NASA/JPL-Caltech; 66 (RT), NASA/JPL- Caltech; 67, NASA/JPL-Caltech/Ames; 68-69, NASA/JPL/ University of Arizona; 70 (UPLE), NASA/JPL-Caltech/ MSSS/LANL; 70 (UPCTR), NASA/JPL-Caltech/MSSS; 70 (UPRT), NASA/JPL-Caltech; 70 (LO), NASA; 72 (UPLE), NASA/JPL-Caltech; 72 (UPRT), NASA/JPL-Caltech; 72 (LO), NASA; 72-73, NASA/JPL-Caltech; 73 (UPLE), NASA/ JPL-Caltech/MSSS; 73 (UPRT), NASA/JPL-Caltech/MSSS; 73 (CTR), NASA/JPL-Caltech/MSSS; 73 (LOLE), NASA/ JPL-Caltech; 73 (LORT), NASA/JPL-Caltech/MSSS.

Chapter 4: 74-75, NASA/JPL/University of Arizona; 76 (UP), NASA/JPL-Caltech; 76 (LO), NASA/JPL-Caltech/ MSSS and PSI; 76-77, NASA/JPL-Caltech/Ken Kremer/ Marco Di Lorenzo; 78, NASA/UC Berkeley; 79 (UPLE), NASA/JPL-Caltech/MSSS; 79 (UPCTR), NASA/JPL/ University of Arizona; 79 (UPRT), NASA/JPL-Caltech/ MSSS; 79 (LO), NASA/JPL-Caltech/University of Arizona; 80 (UP), NASA/JPL-Caltech/University of Arizona; 80 (LO), NASA image created by Jesse Allen, using data from NASA/GSFC/METI/ERSDAC/JAROS, and the U.S./Japan ASTER Science Team; 81, MSSS; 82, NASA/JPL/ASU; 83 (UP), John Hart—*Wisconsin State Journal;* 83 (LO), NASA/JPL/University of Arizona;

84, NASA/JPL/University of Arizona; 85, NASA/JPL/
University of Arizona; 86 (UP), NASA/JPL-Caltech/
MSSS; 86 (CTR), Courtesy Bill Dietrich; 86 (LO), Bill
Dietrich and Marisa Palucis, University of California
Berkeley; 87, NASA; 89, Fred Calef/NASA-JPL; 90, NASA/
JPL-Caltech/MSSS; 92-93, Kenn Brown.

Chapter 5: 94-95, NASA/JPL-Caltech/MSSS; 96 (LE),
NASA/JPL-Caltech; 96 (RT), NASA/JPL-Caltech; 96-97,
NASA/JPL-Caltech; 97 (UPLE), NASA/JPL-Caltech; 97
(UPRT), NASA/JPL-Caltech; 97 (LO), NASA/JPL-Caltech/
Malin Space Science Systems; 98 (UPLE), NASA/JPL-
Caltech/MSSS; 98 (UPRT), NASA/JPL-Caltech/MSSS; 98
(LO), NASA/JPL-Caltech; 99, NASA/JPL-Caltech/MSSS;
100, NASA/JPL-Caltech; 101 (UP), NASA/JPL-Caltech/
MSSS; 101 (CTR), Reuters/Fred Prouser; 101 (LO),
NASA/JPL-Caltech/MSSS; 102, NASA/JPL-Caltech/
MSSS; 103, AP Images/Damian Dovarganes; 104,
NASA/JPL-Caltech; 104-105, NASA/JPL/University
of Arizona; 105 (RT), NASA/JPL/Dan Goods; 105 (LE),
Spencer Lowell/Trunk Archive; 106, NASA/JPL/
University of Arizona; 107 (UPLE), NASA/JPL/Malin
Space Science Systems; 107 (UPCTR), NASA/JPL/Malin
Space Science Systems; 107 (UPRT), ASA/JPL/Malin
Space Science Systems; 107 (LO), JPL/NASA; 108-109,
NASA/JPL/University of Arizona; 110, NASA/JPL-
Caltech/Malin Space Science Systems/Texas A&M
University; 110-111, MSSS; 112 (UP), NASA/JPL-Caltech;
112 (CTR), NASA/JPL-Caltech/MSSS; 112 (LO), NASA/JPL-
Caltech; 112-113, NASA/JPL-Caltech; 113 (UP), NASA/
JPL-Caltech/MSSS; 113 (CTR), NASA/JPL-Caltech/MSSS;
113 (LO), NASA/JPL-Caltech/MSSS.

Chapter 6: 114-115, NASA/JSC/Stanford University;
116, Elisabetta Bonora and Marco Faccin; 117, NASA/
JPL-Caltech/MSSS; 118 (LE), Wikipedia/Benjah; 118 (RT),
Wikipedia/Benjah; 118-119, Mona Lisa Production/
Science Source; 119, NASA/JPL; 120, NASA/JPL-Caltech/
University of Arizona/Texas A&M University; 120-121,
NASA/JPL-Caltech/University of Arizona/Texas A&M
University; 123 (UP), Eckhard Slawik/Science Source;
123 (LO), O. Louis Mazzatenta/National Geographic
Creative; 124, NASA/JPL/University of Arizona; 127
(UP), NASA/JPL-Caltech/MSSS; 127 (CTR), NASA; 127 (LO),
NASA/JPL-Caltech; 128, Chris Gunn/NASA; 129, NASA/JPL-
Caltech/LANL/CNES/IRAP/LPGN/CNRS; 131 (UP), NASA/
JPL-Caltech; 131 (LO), NASA/JPL-Caltech; 132-133, Mark
Thiessen, National Geographic Creative; 133, NASA.

Chapter 7: 134-135, NASA/JPL/JHUAPL/University
of Arizona/Brown University; 136 (LE), NASA/JPL-
Caltech/MSSS; 136 (RT), NASA/JPL-Caltech/MSSS; 137
(UP), NASA/Ames; 137 (LO), NASA/JPL-Caltech; 138
(UP), NASA/JPL-Caltech/MSSS; 138 (CTR), NASA/Paul

E. Alers; 138 (LO), MSSS; 139, NASA/JPL-Caltech/MSSS;
140-141, NASA/JPL-Caltech/University of Arizona;
142, NASA/JPL-Caltech; 143 (UPLE), NASA/JPL-Caltech/
LANL/CNES/IRAP/LPGN/CNRS; 143 (UPCTR), NASA/JPL-
Caltech; 143 (UPRT), NASA/JPL-Caltech/MSSS; 143 (LO),
NASA/JPL-Caltech; 144 (LE), NASA/JPL-Caltech/MSSS/
ASU; 144 (RT), NASA/JPL-Caltech/MSSS; 145, NASA/JPL-
Caltech/MSSS; 146, Dirk Wiersma/Science Source;
147, NASA/JPL-Caltech/MSSS/JHU-APL; 148-149, NASA/
JPL-Caltech/MSSS; 150, NASA/JPL-Caltech/University
of Arizona; 151, AP Photo/Damian Dovarganes; 152-
153, Kenn Brown.

Chapter 8: 154-155, ESA; 156, BAE Systems; 157 (UP),
NASA/JPL-Caltech/MSSS; 157 (CTR), Ben Cichy; 157
(LO), NASA/JPL-Caltech; 158-159, NASA/JPL-Caltech;
159, NASA/JPL-Caltech; 160, NASA/JPL-Caltech; 161,
NASA/JPL-Caltech; 162-163, Christopher Halloran/
Shutterstock; 163, NASA/JPL-Caltech; 165 (UPLE),
NASA/JPL; 165 (UPCTR), NASA; 165 (UPRT), NASA/JPL-
Caltech/Cornell University/Arizona State University;
165 (LO), NASA/JPL-Caltech; 165 (CTR), NASA/JPL-
Caltech; 166 (LE), NASA/JPL; 166 (RT), NASA/JPL; 167,
NASA/JPL-Caltech; 168, NASA/JPL; 169, NASA/JPL-
Caltech; 171 (UP), NASA/JPL-Caltech; 171 (LO), NASA/
JPL-Caltech; 172-173, NG Maps.

Chapter 9: 174-175, NASA/JPL-Caltech; 176, Wayne
Ranney; 177 (UP), NASA; 177 (LO), Reproduced with
permission from the *Annual Review of Earth and
Planetary Sciences,* Volume 42 © 2014 by Annual
Reviews, http://www.annualreviews.org.; 178 (UP),
MSSS; 178 (LO), Chris Hill/National Geographic
Creative; 179, Kris Capraro/JPL; 180-181, Fred Calef/
NASA-JPL; 182 (UP), NASA/JPL-Caltech/MSSS; 182
(CTR), Carolyn Russo/Smithsonian; 182 (LO), NASA/
JPL-Caltech/MSSS; 183, NASA/JPL-Caltech/JHUAPL;
185 (ALL), NASA/JPL/University of Arizona; 186, NASA/
JPL/University of Arizona; 187, NASA/JPL/University
of Arizona; 188-189, Joe Fox/Radharc Images/Alamy;
189, NASA; 190, NASA/JPL/University of Arizona; 191
(LE), MSSS; 191 (RT), NASA/JPL-Caltech/MSSS; 192-193,
Kenn Brown; 193 (UP), David Crisp and the WFPC2
Science Team (Jet Propulsion Laboratory/California
Institute of Technology), and NASA; 193 (LO), NASA/
NOAA/GSFC/Suomi NPP/VIIRS/Norman Kuring.

Chapter 10: 194-195, NASA; 196, AP Photo/Kevork
Djansezian; 196-197, NASA/JPL-Caltech; 198, NASA;
199 (UP), NASA; 199 (LO), NASA; 200 (UP), NASA/JPL-
Caltech/MSSS; 200 (CTR), NASA/JPL-Caltech; 200 (LO),
Mike Nelson/AFP/Getty Images; 203 (UP), NASA/
JPL-Caltech; 203 (LO), NASA; 204, NASA; 205 (UPLE),
ESA/DLR/FU Berlin (G. Neukum); 205 (UPCTR), ESA/

DLR/FU Berlin (G. Neukum); 205 (UPRT), ESA/DLR/FU Berlin (G. Neukum); 205 (LO), NASA; 206-207, NASA; 207 (LE), NASA; 207 (RT), NASA; 209 (UP), AP Photo/NASA; 209 (LO), NASA; 210-211, NASA; 211, NASA; 212-213, NG Maps; 213, NASA/JPL-Caltech.

Chapter 11: 214-215, NASA/JPL-Caltech; 217, Detlev van Ravenswaay/Getty Images; 218, NASA/JPL-Caltech; 219, NASA/JPL-Caltech/ESA/DLR/FU Berlin/MSSS; 220, NASA/JPL-Caltech/Michael Malin; 220-221, txNotAlien/NASA/JPL/MSSS; 222, NASA/JPL-Caltech/MSSS; 222 (INSET), NASA/JPL-Caltech/MSSS; 223, AP Photo/Kevork Djansezian; 225 (UPLE), NASA/JPL-Caltech/University of Arizona; 225 (UPCTR), E.R. Degginger/Science Source; 225 (UPRT), Greg Winston/National Geographic Creative; 225 (LO), NASA/JPL-Caltech; 226, NASA/JPL/University of Arizona and Lori Fenton, Cosmic Diary, SETI Institute; 227, NASA/JPL/University of Arizona and Lori Fenton, Cosmic Diary, SETI Institute; 229 (UP), NASA/JPL-Caltech/MSSS; 229 (CTR), Lance Hayashida/Caltech; 229 (LO), NASA/JPL/University of Arizona; 230-231, NASA/JPL/University of Arizona; 232-233, NG Maps.

Chapter 12: 234-235, Kees Veenenbos; 236 (UP), Time & Life Pictures/NASA/Getty Images; 236 (LOLE), NASA/JPL; 236 (LORT), SSPL via Getty Images; 237, Jet Propulsion Laboratory; 238 (UP), Michael J. Daly/Science Source; 238 (LO), NASA/JSC/Stanford University; 240 (LE), Roberto Barbieri/Science Source; 240 (RT), Jorn Peckmann/Science Source; 241, NASA/JPL-Caltech/Ames; 242 (UPLE), RGB Ventures LLC dba SuperStock/Alamy; 242 (UPCTR), NASA/JPL; 242 (UPRT), NASA/JPL/MSSS; 242 (LO), NASA/JPL-Caltech/University of Arizona; 243, NASA/JPL-Caltech/SwRI; 245 (UPLE), Bill Dietrich and Marisa Palucis, University of California Berkeley; James Bell, Arizona State University; Ryan Anderson, USGS; 245 (UPRT), Brian Hynek/CU-Boulder's Center for Astrobiology/NASA; 245 (LO), Bill Dietrich and Marisa Palucis, University of California Berkeley; James Bell, Arizona State University; Ryan Anderson, USGS; 246 (UP), NASA/JPL-Caltech/MSSS; 246 (CTR), NASA/JPL; 246 (LO), Ron Miller/Black Cat Studios; 248-249, NASA/JPL-Caltech/MSSS/JHU-APL/Brown University; 249, Volker Steger/Science Source; 250-251, NASA/JPL-Caltech; 252-253, NG Maps.

Chapter 13: 254-255, iurii/Shutterstock; 256, Professor Dava Newman: Inventor, Science Engineering; Guillermo Trotti, A.I.A., Trotti and Associates, Inc. (Cambridge, MA): Design; Dainese (Vincenca, Italy): Fabrication; Douglas Sonders: Photography; 256-257, Stocktrek Images/Getty Images; 258-259, Stephan Morrell; 260-261, Nadav

Neuhaus; 263, NASA/SDO; 264 (UP), NASA/JPL-Caltech/MSSS; 264 (CTR), C-SPAN; 264 (LO), Mark Thiessen/National Geographic Creative; 265, ESA/IBMP; 266-267, Rex Features via AP Images; 267, AP Photo/John Bazemore; 268, SpaceX; 269, NASA; 270, ESA/NASA; 270 (INSET), Wikipedia/Artists concept by Nesnad; 272, AP Photo/Mikhail Metzel; 272-273, NASA; 274-275, Kenn Brown.

Chapter 14: 276-277, NASA, J. Bell (Cornell U.) and M. Wolff (SSI); 278-279, NASA/JPL-Caltech/Malin Space Science Systems/MidnightPlanets; 279, Daein Ballard; 280 (UP), NASA/JPL/University of Arizona; 280 (UPCTR), NASA/JPL/University of Arizona; 280 (LOCTR), NASA/JPL/University of Arizona; 280 (LO), NASA/JPL/University of Arizona; 282, NASA/JPL-Caltech/Cornell University/USGS/Modesto Junior College; 283, NASA/JPL/University of Arizona; 284-285, NASA/JPL-Caltech/JHUAPL/MSSS/SETI Institute; 285 (UP), NASA/JPL-Caltech; 285 (LO), NASA/JPL-Caltech; 286, NASA/JPL/University of Arizona; 287 (UP), Mars-One.com/Bryan Versteeg; 287 (LO), Alfred McEwen, USGS Astrogeology; 288-289, NASA/JPL/Arizona State University; 290-291, NG Maps; 292-293, NG Maps.

Original Art: Mars past, present, and future illustrations created by Kenn Brown at Monolithic Studios.

MAP SOURCES

ALL GLOBAL MARS MAPS
Place names: Gazetteer of Planetary Nomenclature, Planetary Geomatics Group of the USGS (United States Geological Survey) Astrogeology Science Center
planetarynames.wr.usgs.gov

IAU (International Astronomical Union)
iau.org

NASA (National Aeronautics and Space Administration)
nasa.gov

Global Mosaic: NASA Mars Global Surveyor; National Geographic Society

MSL/CURIOSITY MISSION MAPS
NASA/JPL-Caltech

Special thanks to Fred J. Calef III, PhD, JPL

MARS ALLUVIAL FANS MAP
Sharon A. Wilson Purdy, Smithsonian Air and Space Museum

INDEX

INDEX

INDEX

MARS UP CLOSE

Marc Kaufman

PUBLISHED BY THE NATIONAL GEOGRAPHIC SOCIETY

John M. Fahey, *Chairman of the Board
and Chief Executive Officer*

Declan Moore, *Executive Vice President; President,
Publishing and Travel*

Melina Gerosa Bellows, *Executive Vice President;
Publisher and Chief Creative Officer, Books, Kids,
and Family*

PREPARED BY THE BOOK DIVISION

Hector Sierra, *Senior Vice President and
General Manager*

Janet Goldstein, *Senior Vice President
and Editorial Director*

Jonathan Halling, *Creative Director*

Marianne R. Koszorus, *Design Director*

Susan Tyler Hitchcock, *Senior Editor*

R. Gary Colbert, *Production Director*

Jennifer A. Thornton, *Director of Managing Editorial*

Susan S. Blair, *Director of Photography*

Meredith C. Wilcox, *Director, Administration
and Rights Clearance*

STAFF FOR THIS BOOK

Andrea Wollitz, *Project Manager*

Marty Ittner, *Art Director*

Nancy Marion, *Illustrations Editor*

Carl Mehler, *Director of Maps*

Matthew Chwastyk, *Cartographer*

Marshall Kiker, *Associate Managing Editor*

Judith Klein, *Production Editor*

Lisa Walker, *Production Manager*

Galen Young, *Rights Clearance Specialist*

Katie Olsen, *Production Design Assistant*

PRODUCTION SERVICES

Phillip L. Schlosser, *Senior Vice President*

Chris Brown, *Vice President, NG Book Manufacturing*

Robert L. Barr, *Manager*

Art Hondros, *Imaging Technician*

The National Geographic Society is one of the world's largest non-profit scientific and educational organizations. Founded in 1888 to "increase and diffuse geographic knowledge," the Society's mission is to inspire people to care about the planet. It reaches more than 400 million people worldwide each month through its official journal, *National Geographic,* and other magazines; National Geographic Channel; television documentaries; music; radio; films; books; DVDs; maps; exhibitions; live events; school publishing programs; interactive media; and merchandise. National Geographic has funded more than 10,000 scientific research, conservation and exploration projects and supports an education program promoting geographic literacy. For more information, visit www.nationalgeographic.com.

For more information, please call 1-800-NGS LINE (647-5463) or write to the following address:
National Geographic Society
1145 17th Street N.W.
Washington, D.C. 20036-4688 U.S.A.

Visit us online at www.nationalgeographic.com

For information about special discounts for bulk purchases, please contact National Geographic Books Special Sales: ngspecsales@ngs.org

For rights or permissions inquiries, please contact National Geographic Books Subsidiary Rights: ngbookrights@ngs.org

LIBRARY OF CONGRESS CATALOGING-IN-PUBLICATION DATA

Kaufman, Marc, author.
Mars up close : inside the Curiosity mission / Marc Kaufman. -- 1st edition.
    p. cm.
Includes bibliographical references and index.
ISBN 978-1-4262-1278-9 (hardcover : alk. paper)
1. Mars (Planet)--Exploration. 2. Curiosity (Spacecraft) I. Title.
QB641.K24 2014
559.9'23--dc23
                2013046521
Printed in China
14/RRDS/1

MARS UP CLOSE was created with generous support from SpaceX. SpaceX designs, manufactures, and launches the world's most advanced rockets and spacecraft. The company was founded in 2002 by Elon Musk to revolutionize space transportation, with the ultimate goal of enabling people to live on other planets.

Today, SpaceX is advancing the boundaries of space technology through its Falcon launch vehicles and Dragon spacecraft. It is the only private company ever to return a spacecraft from low Earth orbit, which it first accomplished in December 2010. The company made history again in May 2012 when its Dragon spacecraft attached to the International Space Station, exchanged cargo payloads, and returned safely to Earth—a technically challenging feat previously accomplished only by governments. Since then Dragon has delivered cargo to and from the space station multiple times, providing regular cargo resupply missions for NASA.

Under a $1.6 billion contract with NASA, SpaceX will fly a total of at least 12 cargo resupply missions to the ISS—and in the near future SpaceX will carry crew as well. Dragon was designed from the outset to carry astronauts; under a $440 million agreement with NASA, SpaceX is making modifications to ready Dragon for crew.

All the while, SpaceX continues to work toward one of its key goals—developing reusable rockets, a feat that will transform space exploration by delivering highly reliable vehicles at radically reduced costs. For more information, visit *www.spacex.com*.